95 Theses *PROVING* Evolution Is A *LIE!*

Fair Use Statement:

In creating this work, I draw upon numerous sources that I have been exposed to over the course of some thirty years of looking into this subject.

Many of these sources I no longer recall the author, nor do I have the pamphlet or video - and so cannot give proper credit in those cases. In other cases, I directly quote or paraphrase published works, some of which are copyrighted works.

I have reviewed the Fair Use information given online by the Copyright Office and firmly believe that my new work falls well within the guidelines offered by that Office.

It is my sincere intention to weave together existing material from others, original thoughts from myself and works where the author has released copyright - to create a wholly new work.

In my mind, I have not violated anyone's copyrights by this work.

Introduction:
Opening the Spiritual 'Iron Curtain'

The priest walked resolutely - hammer in hand - toward the Castle Church in Wittenberg, Germany as he had dozens of times before.

In his other hand was yet another posting he intended to nail to the wooden door of the church for the people to read and discuss.

The hammer fell.

The nail pierced through the document and deep into the ancient door.

The world of men would soon be embroiled in a movement that two and half centuries later would topple the empire that had ruled the earth for a thousand years.

The date was October 31, 1517.

Simultaneously with that hammer blow; in the Spiritual Reality that fully encompasses and permeates this temporal reality of flesh and blood:

A crack formed in the spiritual empire of the Roman Church.

Her time to repent of her sins as prophesied in the Revelation had expired.

Now God would remove her from her position of power.

The process would be long and would involve much bloodshed; but the Reformation would ultimately prevail this time after the efforts of Martin's predecessors - John Wycliffe, John Huss and many others - had been defeated by information suppression and outright violence.

No matter that they did not succeed in their day. The Truth is the Truth no matter if no one understands it, believes it, promotes it or follows it. The Truth cannot be erased nor can it be defeated by tyrants.

Some say that Martin never actually nailed his 95 Theses to the cathedral door and the exact details of the event may never be known, but what has endured are the Truths in the Theses and the debate that followed.

The Christian is defined by:

1) Solus Christus - Christ Alone
2) Sola Gratia - Grace Alone
3) Sola Fides - Faith Alone
4) Coram Deo - man is ever face-to-face before God

At the risk of being too bold - in this 500th year since the release of Martin Luther's 95 Theses - there is at least one more Statement that must be emphasized as essential to believing these other four Doctrines are True.

While a person can certainly 'get saved' without understanding or believing in or being guided by Creation concepts or 'Theory'; if a Christian does not eventually understand and embrace Creationism - they will be at least much less effective or even counter-productive in their Christian witness.

That Statement is:
Sola Creatura - Creation Alone

Specifically, by Young Earth Creation - ALONE - no room for compromise in any way.

Period.

This work will be published on or before October 31, 2017; in the 500th anniversary year of Martin Luther's release of his 95 Theses.

Only God knows what effects it will have in the realm of men or in the Spiritual Reality He governs. It may begin another Reformation of sorts - this time toppling another corrupt SPIRITUAL empire - the empire known as the Religion of Evolution that has reigned for the last fifty or so years.

Other brave people like the Drs. Henry Morris (sr. & jr.), Ken Ham and Kent Hovind - and many, many more - have fought to overthrow this empire that replaces the Truth of the Bible with the musings of men.

Like Huss and Wycliffe, their work has had an effect, though not yet the full effect of toppling the towers of 'so-called science' ruling the thoughts of too many today.

These mere human promoters of Evolution do this for the express (but unstated) purpose of promoting the idea that they have evolved into gods themselves; therefore the laws that apply to the rest of humanity do not apply to them.

Is the Religion of Evolution really so terrible that I should draw a parallel between it and the Roman Church taken down by the Reformation? Consider these facts before you reject the thought.

Evolution is the ROOT concept behind:

- Racism: Without more than one race, there can be no Racism. If we are ALL 'cousins' descended from a single pair of CREATED human beings as God tells us in the Book of Acts: We are all of 'one blood'. If we all followed the Creator's remedy of 'love thy neighbor as thyself' - world peace WOULD be more than just a phrase to be uttered by the finalists of beauty contests.

- Slavery: This off-shoot of Racism exists because Evolution claims one race 'superior' to another race - giving 'justification' for the 'superior' race to subjugate the 'inferior' race.

- Sexism: Charles Darwin believed women inferior to men - so why let them vote or hold public office? Many cultures treat women like cattle or property. God created woman from man's body - therefore women are intrinsically equal to men in the eyes of the Creator.

- Abortion: More than 100 million human babies worldwide have been murdered by now. Why? Evolution DENYS the humanity of the unborn; proclaiming these pre-born human beings with their own unique and fully human DNA as only a 'blob of tissue' or 'at this stage a fish'.

- War: When fought in the name of Evolution like WWII was; those killed by the 'superior' force are NOT equal before God; not our very 'cousins'. This lessens the consequence of killing them in the mind of such men. Adolph Hitler had a list of lesser evolved humans that had to be eliminated for his 'master race' to be fully realized.

- Atheism: The root belief system of Communism and all it's off-shoots. Most of the major wars of the last 200 years have been started by tyrants and tyrannical governments that ascribe to Atheism and Evolution.

- Homosexuality (and all other perversions of human sexuality): These are off-shoots of Atheism's removal of Creator God from the thoughts of mankind. Humans now believe they can 'identify' as anything they 'want to be' to 'fully realize who they are'.

In all, the Religion of Evolution has the blood of many hundreds of millions of humans and the cruel suffering of many hundreds of millions more on its hands. Yes - Evolution MUST come crashing down someday just as the corrupt Roman Church of Martin Luther's day that had slaughtered tens of millions HAD TO be destroyed.

The Religion of Evolution controls nearly all the colleges - sadly, even some Christian ones - being taught as if it were not a religion at all, but as if it were actual proved science. Millions upon millions of people for more than two generations in this country have had this dogma drilled into their brains. They have been taught by their teachers and professors to laugh at and ridicule those who identify Evolution for what it actually is: A bald faced LIE.

Yet, when an Evolutionist is pressed to give a SINGLE, UNDISPUTED piece of evidence proving Evolution to be actually, objectively true - they simply regurgitate what they were told: There are mountains of evidence!

When pressed further to give their single, best evidence - they often realize that they actually have none - and then devolve into name calling and worse. In this work, we will scale their 'mountain of evidence for Evolution' in less than a single step - for it is actually no mountain at all and not even a mole hill - in reality not even an ant hill.

If Evolution be actually true; it should be possible to have a rational conversation laying out proof so convincing that any reasonable person would agree. Yet the proponents of Evolution cannot do so.

I have engaged thousands of them for more than 30 years in all sorts of discussions online, in person and by letter. In all that time NOT EVEN ONE believer in the Evolution Religion could give a SINGLE, UNDISPUTED evidence showing Evolution was a correct concept.

Sadly, even many or most Churches have ceded the battle field to self proclaimed 'scientists' and 'experts'; allowing the Bible to fall into disrepute. These pastors then either ignore the topic or even tell their flocks that Genesis is something other than the Historical Truth.

They may not say this outright, but by not being willing or able to refute the Religion of Evolution's claim of Atheists that 'Genesis 1-11 is not literally True' - they undermine nearly every doctrine in Scripture!

These 95 Theses that follow - each one individually - if ANY are correct - show Evolution is a Lie that must be totally rejected as having no merit whatsoever. Just like a single pin out of a box of pins is capable of destroying an inflated balloon; EACH ONE of these Theses - by itself - shows Evolution to NOT be science, but actually a FALSE RELIGION.

I originally added this list of 95 Theses to the backs of my other books (Lumpy Oatmeal with Raisins and Cinnamon! and Behold Now Behemoth: Dinosaurs All Over the Bible!) so that the readers of those books would have some 'ammunition' when discussing the topic with others.

Now God has laid it on my heart to expand those short Theses into a stand-alone book where each will be treated more thoroughly. Be assured, EACH ONE of these Theses is a sound reason that Evolution is a completely wrong interpretation of the available evidence.

Some Theses will be logical arguments; some will be from science. Some will be examples observed in our world that can ONLY be explained if this earth and everything in it was in fact CREATED by Almighty Creator God.

I tested these 95 Theses by posting them online as You Tube videos and then venturing into Atheist and Evolutionist owned website chatrooms to let them try to tear them apart.

For more than a year of intense debate with these folks; NOT EVEN ONE of the Theses were overturned or shown to be wrong by the hundreds of people who challenged them! A couple of the Theses were shown to need stronger examples to demonstrate the Theses, but NONE were shown to be wrong in the basic premise put forth.

The reader of these Theses can be rest assured that using them will render your opponents powerless. Because of this, get ready for them to pull out the only weapons they have left: Ridicule, scorn and the worst 'cuss-rants' you have ever heard. But that is how you know they have no answer to your argument and that you have soundly beaten them!

Each Theses will have its own section (some may be combined if closely related) and each will be a page or so. In that space; I will state the concept and discuss it logically from both the A-theist (no-God) view and the Theist (Creator God) view as applicable. The reader then will be able to clearly see which view actually aligns with the real, observable evidence and which view does not.

These 95 Theses will show clearly that in EVERY SINGLE CASE, the correct view will be the one that indicates there actually is an All-Mighty, Creator God! I would suggest that since this is True, it would be a very good idea to find out about Him from the Book He Authored (that would be the Bible), figure out what He wants from you (since you are His since He created you) and then DO what He says (get saved and then serve Him).

May God richly bless you as you come to understand that the Bible is His-Story - real, actual HISTORY - not just some collection of camp-fire tales. There REALLY IS a Creator God - and this Universe with all its vastness and glory is only a TOKEN of His might brought into existence by His speaking a few Words!

Note: For most of these Theses, I did a general review of Evolution friendly websites - to see what they say about the issue. Sometimes I mention a particular website, most of the time not as the general sense from many sites was used. I did this because one major criticism leveled against Creationists is that we simply parrot each other. Therefore, I will use their OWN words and musings showing how utterly they fail to make their case.

As I have been doing this for more than 30 years and found all of this material in many places - it is impossible for me to credit accurately where each thing I cite came from.

Many of the Theses are in total or in part my own reasonings. The sources for the rest of the Theses would have come from the following organization's publications or people's published materials - used or summarized by the Fair Use guidelines of the United States Copyright Office:

Answers In Genesis led by Ken Ham

Institute For Creation Research led by Dr. Morris

Dr. Kent Hovind

Dr. Carl Baugh

Dr. Walt Brown

The Revised Quote Book by the Creation Science Foundation

and so many others that I cannot recall by name.

Truly, I am standing on the tall shoulders of these pioneers and tireless people; shouting out the Truth as Wisdom herself is said to do in the Proverbs.

May the masses listen this time and may the new Reformation begin!

Theses One and Two:
The Law of Conservation of Energy and the Law of Conservation of Mass

Premise: Mass and Energy exist and all known things consist of some combination of mass and energy. Both had to have come from somewhere or SomeOne - for any five year old knows that something cannot come from nothing.

There are only three possibilities for how everything in the Universe got here and each requires a person to believe something SOLELY on FAITH alone.

Two possibilities are 'no-God' or A-theist scenarios and the third posits that there is a God, making it a Theist scenario. ALL three are RELIGIOUS postulations in that EACH can only be believed or disbelieved. Regardless which one a person chooses, that person is being 100% religious in their choice.

It does not matter if a person is a Creationist or Evolutionist - ALL agree that the observable Universe is governed by forces and principles that man has found no demonstrable exception to. After much discussion; these unwavering principles have been given a special name: A Natural Law or a Law of Nature. Sometimes a Natural Law may also be expressed as a mathematical statement and other times a logical statement will be used to describe it.

The Law of Conservation of Energy and Mass is usually stated as follows:

Energy / Mass can neither be created nor destroyed, but only altered in form.

I say it this way, "Energy /Mass can neither be created nor destroyed (by any process known to MAN) but can be altered in form either by natural means or by man-made methods.

This principle is provable by human experiments using 'closed systems' - which is a system that has known boundaries. Since ALL of man's science involves these 'closed systems' - an assumption MUST be made regarding ALL systems - closed or open - up to and including the known Universe itself - which; if there is a God that Created it, even the Universe would be a closed system to Him.

Because man MUST use a relatively small, closed system to study - we are forced to assume the principle that holds true in that system will at least generally hold true throughout the Universe.

The Atheist will assume - because it is the only way for his no-God scenario to have any chance at all - that as soon as one leaves the closed system of an earth-bound experiment; the principle completely breaks down in at least isolated places and therefore has no merit. This allows their fertile imaginations to suppose that somehow (though never exactly explaining how) an entire universe can simply pop into existence (see National Geographic October 1999 issue).

In this way, the A-theist exercises far MORE religious faith than the Creationist does! His 'faith alone' or 'sola fides' is in NOTHING becoming something by some unknown mechanism!

Logically, then: Since no one has ever observed NEW (not previously existing) Energy or Mass appear out of nowhere or nothing-ness; man LOGICALLY ASSUMES no new mass or energy is coming into being.

THIS MEANS: Since there IS both mass and energy now in existence - it HAD TO have come from somewhere or SomeOne. There logically had to have been a beginning or else some 'self-existent' Being or force gave rise to the universe. Thus, the universe's start was a one-time event that no human being witnessed. It follows then that man - who came about later - can only attempt to offer explanations that fit this Natural Law.

The A-theist assumes one of two concepts for the origin of the universe - since he believes there is no God. Both VIOLATE the Law just stated! Here is how the A-theist's arguments fail.

He believes that the Universe just somehow came into existence from a 'state of nothing-ness' for no reason and having no purpose whatsoever. This is the majority view of Atheists - at least as far as my experience with them indicates.

Alternatively, the minority of A-theists may believe that the universe itself is self-existent and our existence is just the latest version of that eternal self-existence.

Note that both views are 'sola fides or faith alone' statements of belief and can never be proved in a laboratory. Both views FAIL to identify a plausible CAUSE for which the Universe's existence is the EFFECT.

The Creation believer assumes that there had to be some ultimate Source for all of the energy and mass we observe in the universe. The Creation believer - like the A-theist - chooses to believe something on faith alone. However, UNLIKE the Atheist, he concludes that the best explanation for the universe's energy and mass is a view that agrees with the known Natural Laws of Conservation of Energy / Mass: Creator God or an Intelligent Designer if he is not willing to say the word 'God' brought the Universe into existence.

Notice:

Both the Creation believer and the Atheist believer are BELIEVERS of some kind of story.

BOTH make assumptions that can only be believed!

BOTH are TOTALLY religious when they do this - but being 'religious' bothers the Atheist; so he claims he is being 'scientific' by believing what he believes on faith alone.

This is the Atheist's first lie - from which all others proceed - for his belief has no - nor can it have any - basis in experimental science.

The Creation-ist is intrigued by the thought of a Creator and so goes looking for an origin account. He searches all over the earth to see if any match what experimental science in closed systems indicate (the Natural Law) and logic (that there had to be a Creator) tell him is True.

He finds one particular account - the only one in the world - that begins this way:

"In the beginning, God created the Heavens and the Earth…."

This account aligns perfectly with the Laws of Conservation of Energy and Mass! It identifies a sufficient Source for all the vast amounts of energy and mass we know to exist.

It matches the logical evidence and the Natural Law far better than the 'no-God' account that the A-theist ascribes to - which does not match either logic or Natural Law.

Many millions of people choose to believe the Bible's opening statement; whether or not they realize that there is a Natural Law that fully supports it.

Summary:
Energy and Mass DO exist.

Logically they had to have come from somewhere or SomeOne.

The best explanation is a Creator.

The best, most comprehensive Creator account is found in the Bible.

Atheists and Evolutionists must believe in one of the other two possibilities for the existence of the Universe - and they must believe it on faith alone.

Both of these possibilities violate the Laws of Conservation of Energy and Conservation of Mass.

These two Theses completely destroy Evolution all by themselves - for if the most foundational doctrine of the Atheist and Evolutionist cannot be True:

NOTHING ELSE that is build on their foundation of shifting sand can be fully or even partially True.

They may occasionally say something that is True during their musing - hey, even a blind squirrel finds a nut once in a while - but if they do say something that is True, the Creationist will agree with that singular point anyway!

Since the war is already won with just this opening salvo: The rest of the Theses are the 'mopping up' of the pockets of resistance.

Many Evolutionists demand that they get to start with something already existing without having a valid concept of how that something came to be in the first place. To prove that their view is but assumptions built on lies built on supposition - all of which collapses when their foundation of sand is exposed: We will entertain their fantasy in future Theses and let them start with something that either came from nothing or has always existed with no explanation of how this could be.

The old joke about the Atheist arguing with God applies to this situation.

It seems that one day an Atheist wanted to prove to God that He was not necessary - by showing that life could occur from just a pile of dirt and water to make a dead chemical soup that would 'come alive'.

The Atheist said to God, "Now give me some dirt and water so I can show You how it can happen…."

To which God replied, "Make your own!"

Checkmate.

Theses Three:
The Law of Increasing Entropy

Premise: In every system, disorder will increase unless intelligent intervention causes order to remain constant or increase.

Also known as the "No Free Lunch" Law - this Law is usually stated something like this: Disorder rises in every closed system. Other ways it may be stated:

"Everything rusts, busts, dies or decays over time"

"The Universe will someday run down and experience a heat death"

If a person thinks about this a little, there are some LOGICAL conclusions that can be drawn.

First: Since the Universe is known to be 'running down' - there logically MUST have been a fully wound-up state or a 'beginning'.

Since this is logically True - the Atheist that believes the Universe has simply always existed is now shown to be violating THREE Natural Laws. The Atheist that claims the entire Universe just popped into existence out of absolutely NOTHING is also shown to be WRONG; since this Law rests on the Conservation of Energy / Mass Laws as its base assumption.

One might foolishly entertain the idea that a single sub-atomic particle might 'pop out of nothingness' as claimed in the National Geographic October 1999 issue, but the ENTIRE UNIVERSE popping into existence without a God to be the Source of it? Seriously?

Someone was on a really bad LSD trip when they came up with that one!

Is it any wonder the Evolutionist/Atheist affectionately refers to their 'Higgs-boson' as the 'God-particle'! The 'reason' they claim this is possible is again because the Law can only be demonstrated in closed systems here on Earth.

Even though they cannot demonstrate either of their claims can happen to result in a Universe, they say one of the two did happen anyway - because we are not dealing with a closed system!

Therefore they conclude (wrongly) that this Law does not apply at all to the general case or to open systems!

A review of their websites indicates that they believe that if 'pockets of high entropy' escape from the open system at a rate faster than entropy is rising in the system - the net result will be a more ordered system!

My head hurts just trying to follow their 'logic'.

As will be shown by many of the Theses in this work, Evolutionists make MANY unproved and unprovable claims - and then treat these claims as if they were established fact not open to challenge at all.

In science, where a Law is demonstrable on a small scale in a closed system - the **GENERAL PRINCIPLE** being studied is **ALWAYS** the **TENDENCY** even when considered for the open system case. The math may not work out as precisely for the open system case, but to claim you can arrive at a totally different answer is NOT sound reasoning. It is a CLAIM, nothing more.

Claims made by professional scientists are STILL just claims. Just because they have strings of letters after their names DOES NOT make their claim True. A thinking person must still consider logically if the stated claim makes any sense or could possibly be True.

Remember, to God - even the entire Universe is a 'Closed System' - therefore, to the Law Giver (God); the Law DOES apply throughout His entire Creation. Fallible men desiring to be god deny all logic so that their vain hope of taking God's job one day may have a chance.

Does the Bible hold an account of mere man desiring to be his own god?

Yes.

Genesis chapter 3 tells of the day the first man and woman foolishly listened to Nachash - the hissing one - the serpent from the Garden. Satan apparently had possessed this creature and spoke to them through it - first casting doubt on what Creator God had told them; then telling them that they could become as gods themselves.

Atheists /Evolutionists poo-poo the idea of talking animals; yet a Creator could easily make it possible for animals to speak - since the ALL-Mighty can do anything.

It is NOT that animals speaking is known to be impossible - it IS that a talking animal WOULD indicate that the Bible is True in this regard.

I find it amusing that Atheists / Evolutionists spend hundreds of hours trying to get chimps to speak and write - so that they can claim we evolved from the common ancestor of man and chimp! Why are THEY not ridiculed for this belief that they will one day succeed in getting a chimp to talk?

The reason is because these folks DO NOT WANT there to be a Creator God. If there is a God, the job they want is taken - permanently! It would also mean THEY are created beings - as such, they are under the Authority and Rule of this Creator forever. This chaps their hide - a lot. I tell them to get some spiritual Vaseline - because they are going to need it!

We can safely say that these Atheists are WRONG in their most basic premise. Why then should be believe ANYTHING they build on their false premise? Obviously, we should not believe anything they say.

Molecules to man Evolution is a construct build squarely on Atheistic false premises. If you 'pull the string' and follow their 'so-called science' backward to their origin statement - we find that BOTH of their foundational no-God belief concepts are completely WRONG and have NO possibility whatsoever of actually being correct in the slightest degree.

I could stop this book right here after just three Theses! Actually, I could have stopped after the first two!

Remember as you continue reading that it only takes a single pin to pop a balloon.

However, we will continue to nail the coffin shut on the rotting corpse of the Religion of Evolution by adding another 92 Theses.

We could add many more than 92 - for literally EVERY SINGLE actually True fact will be found to align perfectly with a Creation viewpoint while at the same time utterly proving Evolution to be false.

Theses Four:
The Law of Biogenesis

Premise: Life only comes from pre-existing life and cannot arise from non-living materials without a Giver of Life.

This Law has been called the most proved Law ever!

When Charles Darwin's book came out - some of his proponents immediately proclaimed there to be a 'Law of Spontaneous Generation' - i.e. life could arise from non-living matter on its own. Many others had proposed similar things for many hundreds of years, but now the chorus reached a fevered pitch.

The Atheist's experiments took water ***originally open to the air***, then sealed off the air and boiled it for various lengths of time to sterilize any pre-existing life. In each case, microscopic organisms appeared in short order. They concluded that obviously, life CAN come from just water!

Not so fast….

Louis Pasteur famously showed this simplistic experiment to be utterly false more than 130 years ago. He did this by repeating their experiment - but doing a much better job of ensuring that DUST could not enter the 'sealed' jars.

His sealed jars remained free of any life for many months. When some complained he had destroyed the very ability of the water to generate life, he opened the flasks and soon organisms appeared.

It was therefore shown that these microorganisms must have hitched a ride on the dust in the air and some had the ability to survive the harsh environment of boiling water. We now know of organisms that actually thrive in such extreme environments.

Louis then proposed a different principle that matched the experimental data better. This principle would go on to become the Law of Biogenesis (Bio = life; Genesis = beginning of). This Law could be stated as the Law of Life's Beginning: Life from life - every time; without known exception.

Yet Atheists poo-poo this Law as well - going so far as to say it is not a Law! Because their false religious construct REQUIRES life to come from non-living chemical soup; they have no choice but to CLAIM without proof that this Law is not a Law. They then spend countless hours and mountains of money trying in vain to find some way, any way that life could have come about without Creator God.

Though they still have NO PROOF their idea could ever happen even after more than 100 years of trying; THEIR idea is taught as fact in virtually every 'science' class in the world!

My daughter brought home her textbook from 7th grade science class. On page 222 was Louis Pasteur's work proving life could never come from non-living sources. On page 223 the very next section was titled, "How Life Arose On The Early Earth"! The next pages went on to say that the earth cooled, it rained on the rocks for millions of years, made chemical soup and the soup 'came alive' - i.e. spontaneous generation.

Page 222 said rightly that spontaneous generation could not happen and that this had been proved by Louis Pasteur. Page 223 said '…but it happened anyway on earth…' - and this is called SCIENCE! I want my money back!!

Saying that life arose from non-life is nothing but a CLAIM; originating from a RELIGIOUS construct of the unobserved past!

Why did they say this? - other than because there is no way for molecules-to-man Evolution to work unless spontaneous generation happened at least one time.

They said this because some CAREFULLY DESIGNED and INTELLIGENTLY EXECUTED experiments in a man-created laboratory under carefully controlled conditions using some chemicals and an energy source succeeded in making basic amino acids which are termed 'the building blocks of life'. We will look at these experiments more closely in later Theses.

What IS WITHHELD from the students is that these experiments produced equal numbers of 'left-handed' and 'right-handed' amino acids.

Life ONLY uses left-handed amino acids. If even a single right-handed amino acid is used, life is not possible.

What these experiments actually proved is that UNLESS there is a creator (small 'c' - i.e. human creator) and intelligent processes not realistically demonstrated to exist in a natural setting: NOT EVEN the proper basic building blocks of life will form from dead chemicals!

I have argued on-line with many Evolution believing persons. ALL take one of two tracks:

Some want to start with a pre-existing super-cell to evolve all other plant and animal life from - without ever telling us how this super-cell came to be.

The rest simply dismiss this most proved Law without any proof at all that some counter method of life is possible. They say this is reasonable because men will one day figure out how it happened.

Let me get this straight: The most intelligent men using the most sophisticated computers and laboratory equipment STILL cannot figure out how brainless, purposeless, random chance managed to put together that first super-cell; yet it is somehow 'settled science' that it did happen?

These men have no shortage of faith. It is just that their faith is completely MISPLACED in fallible, created men rather than the Self Existent Eternal Creator God.

Theses Five:
The Laws of Motion and The Law of Conservation of Angular Momentum

Premise: The motion of objects obeys set principles with unerring constancy. If we see object(s) moving in ways that contradict these known principles - we must propose a valid reason WHY they are moving in this way.

In our world, we often encounter objects that are moving. We also see objects that are not moving. If an object is not moving, we understand by common experience that some kind of force is required to get the object to move - or if the object is moving, a force must be applied to stop it from moving or change the direction of its motion. Scientist or not, able to do the math behind it or not - we ALL KNOW these things.

In the vacuum of outer space, there is one thing missing that is present here on earth: Friction. In outer space there is nothing to push or pull against. If an object were motionless in outer space, it would remain this way until some force acted upon it. The object would then move in direct relation to the force that acted upon it and have some momentum.

The object would move according to the amount of force and the angle with which the force was applied. Therefore, if we study the motion of the object, we can logically figure out many things about the force that caused the object to start moving. Because in outer space there is no friction, two Laws apply.

The Law of Motion basically says that a force must be applied to get the object at rest into motion. The Law of Angular Momentum basically says that the object's motion must have some relationship to the force that caused the motion. There are mathematical equations that can be done - but will not be done in this book. Look on-line if you care to at the many sites that will help you understand the math.

We can give an analogy of this using a merry-go-round. Load some kids on it and apply a force to get it moving. Apply more force and it will go around faster. Keep doing this until the kids start to lose their grip and fly off! Study the motion of the kid that flies off and you will see that his motion bears a direct relationship to that of the merry-go-round.

Atheists / Evolutionists have a grand fairy-tale regarding the origin of the Universe from some 'Big Bang'. No human witnessed this event, so all we can do is look into outer space at the objects supposedly resulting from the Big Bang. We see the objects moving. We understand from math and science that the motion of these objects MUST bear some relationship to the event that is claimed to have set them in motion.

If we observe consistent motion of these objects, we could conclude that they all were set in motion by the same event. The trouble is - there are demonstrable cases where very large objects called planets are spinning on their axis in the wrong direction and / or are orbiting around something in the wrong direction if Big Bang were true! Whole planets, moons and even our Sun do not move in accordance with the Law of Conservation of Angular Momentum!

A logical conclusion then is that the Big Bang is an INCORRECT construct of the unobserved (by man) past. We can be sure this event as described to us by Atheists / Evolutionists is WRONG.

Is there an explanation that COULD BE right? Yes! "In the beginning God Created the Heavens and the Earth…." This account goes on to say that Creator God made the stars, planets and moons, etc. - He made everything!

He could have spun the planets and set the orbits of the moons ANY way He wanted to! He may have done so at the odd angles we see just to show Atheists / Evolutionists that He exists - knowing that one day man would invent telescopes and spaceships - travel to other planets and our own moon. In doing so, man would see that these two Laws are not obeyed by the motion of many objects and so logically there must be a different answer than a God-less tale for the existence of these objects.

Folks - hear me: It is LOGICAL to believe in Creator God! The Bible is completely backed up by TRUE SCIENCE! It is the Atheists / Evolutionists that must exercise more religious faith in their concepts than we have to as Christian believers in Creator God and the Bible!

There is no need to fear these people who confidently and arrogantly say that you are foolish for believing in Creator God. They have been brainwashed by professors in college who were themselves brainwashed by their professors.

Do not get angry at them - pity them! They NEED YOU to become able to defend your faith so that they one day may come to faith in Creator God and Jesus Christ His Son.

Theses Six:
The Law of Gravity

Premise: Given the well-defined structures we see in outer space, we can conclude that Gravity has not been acting for many millions of years - but rather only a few thousands of years.

Evolution needs time - and lots of it - to make the claim seem possible that amoebas can turn into men if only given enough time. Time is actually Evolution's foe; even if there was millions of years of it. The question becomes; is there enough time for a molecule to become a mouse and then a man?

What are some measures of time aside from man-made watches?

Gravity is a relatively weak force, yet acts throughout our Universe such that structures like galaxies and solar systems are held together - loosely. Because the gravitational force is so weak, we must ask ourselves if it is logical to believe that these structures could have held together in such an ordered state for Evolution's proposed billions of years.

The standard math equation for gravity has a term in it: Mass. The mass of the objects in question bears directly on the amount of gravitational force that is felt.

One clue that Atheists / Evolutionists KNOW that gravitational calculations do NOT provide them with enough time for worms to evolve into even salamanders is that they desperately try to find more mass in the Universe. They do this to allow there to be enough time, for they know without this extra mass - their 'just so' story of an ancient Universe is simply not feasible.

Enter two new imaginary characters: Black Holes and Dark Energy!

Atheist astronomers have been looking for direct evidence of Black Holes for more than 50 years and still have found none. Many websites CLAIM that there is direct evidence that they exist, but if you read the articles - one finds their 'proof' consisting of mere claims or statements that the men writing can think of no other explanation for what they believe they see. Therefore, Black Holes remain just an imagined construct put forth in a vain attempt to save the hopelessly flawed idea that this Universe is some 20 billion years old.

Atheist astronomers know they should not see tightly packed galaxies UNLESS there was A LOT more mass out there than we can detect. These folks had to 'find' the missing mass or admit that the age of the Universe as shown by the standard calculations is on the order of thousands of years, not billions.

This missing mass HAD TO BE invisible because they could not see it even with their most powerful telescopes.

The story goes that Black Holes are old, dead stars that have collapsed in on themselves and now suck in everything - even light!

Well now! If this were True - just how many of these cosmic vacuum cleaners would there have to be? Hmmm…. each solar system would have to contain several….. something like ten Black Holes for every star we can actually see….

Hmmm.... each solar system has ten of them - sucking up everything for many billions of years..... Hmmm... just exactly how is it that WE are still here and have not been sucked up yet?

Some twenty years ago, an article in Smithsonian admitted little hope of ever confirming Black Holes and so proclaimed Dark Energy as the 'missing mass'! Dark Energy was not concentrated mass like in a dead star, but rather invisible 'packets of energy' so small and so undetectable that they are swarming all around us and even travelled through us without our being aware of them!

How convenient - cannot see Dark Energy or detect it - but it is certainly known to be there and it is responsible for holding the whole Universe together! Really? Are these folks smoking something?

If you cannot see it or demonstrate it - you are welcome to **RELIGIOUSLY BELIEVE** it exists - but **DO NOT** call it science! Evolutionists loudly mock Christians for believing in a God we cannot see! They say we believe in magic. I say it this way - BOTH sides may well 'believe in magic' - but at least the Creationist side has a Magician to DO the magic! Both views of the Atheist / Evolutionist are 'magician-less' constructs!!

These folks have no trouble believing in unseen realities UNLESS it involves an unseen Creator God to whom they will have to give an account for their lives one day.

Again, billions upon billions of dollars of **YOUR TAX MONEY** have been spent trying to find the missing mass - all of it a waste.

Well, maybe not a total waste - this intellectual welfare program has kept a lot of PhDs off the street corners begging for money holding cardboard signs saying, 'Will **THEORIZE** for food!'

What of Creation Theory and its mere 6000 years of His-Story. Hmmmm….. fits **PERFECTLY** with the known evidence from the observable Universe.

"What can be known of God is clearly shown by His **CREATION**, being seen…." - where have I heard this before? Oh, yes! The Bible! It is not a book of fairy tales.

The ones making up fictional stories are the so-called scientists and self-proclaimed experts. Their research grants depend on them coming up with ever more imaginative God-less explanations that will require billions more dollars and decades more wasted research. They need to get a real job.

Theses Seven:
The Law of Electromagnetism

Premises: So far as man knows; set materials / conditions are required for an electromagnetic field to exist. There is also a known subset Law of Electromagnetism that states: Opposite Charges Attract and Like Charges Repel.

Theses 7: The Earth has a magnetic field. Where did it come from? No one knows for sure.

Mercury, Earth, Jupiter, Saturn, Uranus and Neptune have magnetic fields, Venus and Mars do not. Why? One would think that if this solar system did in fact form from the same events and materials; ALL the planets would have formed under similar circumstances and have similar characteristics.

Yet, the planets in our solar system are very different from each other in many respects, including the presence or absence of an electromagnetic field.

From experimentation we have proved that special circumstances and materials are required for such a field to exist. This field is critical for cell reproduction, animal migrations and to keep the Sun's lethal radiation sufficiently far away from the surface of the Earth. In other words: There would be no life on Earth if this field did not exist exactly as it does.

Evolutionists make claims about the nature of the core of the Earth, which they ASSUME must be of a certain composition so that this field could exist; but they have never dug even through the crust of the Earth - so they do not know how this field is actually generated.

Creationists have their own theories as well, but we realize there are some things that are and may always be knowable ONLY to the One Who Created this Earth - and we are OK with essentially crediting the Creator with installing this feature in the only known planet that harbors life.

The point of this Theses is that Evolutionists once again confidently state as settled fact something they can ONLY theorize about.

What we DO know is that this field is decaying measurably over the last 70 years. If one extrapolates the measured decay rate backwards in time - we find that the field would become too strong for life to exist just a few tens of thousands of years ago. This is yet another evidence from actual science that indicates the Earth is not billions of years old, but only thousands - and that means there is not enough time for Evolution to allow 'lesser' creatures to evolve into 'greater' creatures.

By the way, WHY are the critters on Earth called creatures?

If Evolution is actually True, should they not be called 'Evolutures'?

The very term CREAT-ures indicates that they obviously were CREATED.

Electromagnetic fields also include the Electrostatic Law of Attraction and Repulsion. This is important to understand because our experiments at the atomic level indicate there are negatively charged electrons orbiting positively charged protons in the nucleus.

This begs the question. Evolutionists would have us believe that this structure and behavior of the electrons orbiting atoms have been in stable orbits for up to 20 billion years!

It is a miracle that the electrons have not had their orbits decay into the positively charged protons for the few thousand years that Creationists say they have existed!

We will speak more about atomic structure in later Theses. For now, it is enough to understand that yet another Natural Law tells us in no uncertain terms that crediting the Creator regarding certain aspects of His Creation is not just 'Inserting God into the Gaps' of our knowledge - it is the logical and well reasoned choice - far more sound a choice than what the Evolutionists choose to believe.

The only difference is that Creationists tell you when they have run into a currently unknowable aspect of our existence and begin to claim that God was involved. Evolutionists simply drive right on into speculation and claim their musings are proved facts.

To the Atheist - since there is no higher Authority - whatever dribbles out of their mouths is just as factual as something that can be proved in the laboratory; that is until their fairytale is demonstrated to be in violation of proved Law or proved wrong by some experiment.

When proved wrong, oftentimes the Evolutionist will simply ignore the new evidence and even continue to teach unsuspecting students known wrong information!

Biogenesis proved Spontaneous Generation wrong in the 1800s; yet my daughter's science textbook told her it happened at least once on Earth - and this was taught to her in the late 1990s!

Let us move on to the next Theses.

Theses Eight:
The Law of Chemical Concentration and Dilution in Water

Premise: A chemical or solid that does not evaporate - when dissolved in water - will rise in concentration as the water is removed through evaporation. Further, if a concentration / dilution process has not reached equilibrium; this process has not been occurring for a long enough time to reach equilibrium.

Theses 8: Chemistry in general - by itself - could provide many more than 95 Theses to show Evolution could never have gotten off the ground by forming even a first cell of any sort, let alone a Super Amoebae that magically turned into bananas and baboons. This book will touch on the topic of Chemistry in only a few ways.

The wonderful part about chemistry is that numerous equations are available. These equations can be rearranged by algebra to determine some value other than the one the equation was originally made to find.

In the case I cite here, we use the general Chemical Concentration and Dilution equation - designed to usually find the Final Concentration after a period of time has elapsed. Through algebra, we can rearrange the equation to determine how long a process has been occurring.

When we solve for 'how long' - we get some very intriguing results! Here is the general equation in words rather than symbols:

Final Concentration = Initial Concentration + Addition Rate - Removal Rate

The two rate factors have time included in them - and if we solve the equation to find a value for time and see the result is way off from what Evolution claims and needs - we can be sure Evolution is WRONG.

Various websites have done these calculations. The Institute for Creation Research (ICR) site tells us that at today's slow rate of buildup, even if the oceans were completely salt free at the beginning - the maximum age of the oceans is only tens of millions of years old.

The ICR site correctly says the oceans continue to get saltier - using actual research data for their claim. Another site called Gizmodo; says all the salt being deposited today ends up on the sea floor - but this is chemically impossible and could not ever be independently determined anyway. It is just a claim so that Evolutionists can poo-poo a powerful indicator of a YOUNG Earth.

Evolutionists need lots of time for their imaginary process to have any chance of fooling anyone into believing it - so they claim their Super Amoebae came to be in a 'warm, salty ocean eddy or tidal pond' a couple billions of years ago.

Hmmm…. is this a reasonable belief? Could it possibly be True? Let's use the above concepts to find out!

If you go to the ocean right now and take a sample of the water and analyze it for it's concentration of salt; you will find about 35 parts of salt per thousand parts of water. That is a lot of salt! Each part per thousand equals just under about four million tons of salt in each cubic mile of seawater. This is the current or 'Final Concentration' for our investigation.

It cannot be known if the oceans were ever completely free of salt - but since we want to know the maximum time this process could have been occurring - we will assume the Initial Concentration of salt was zero ppm.

We know from observation and experimentation that salt goes into the ocean and leaves it at known rates today. Salt is added by runoff from the land via rivers and leaves via salt spray and other mechanisms. If this process has been going on for a couple billion years, the oceans would have reached equilibrium concentration by now - hence the false claim put out by Gizmodo.com. The only problem is - the oceans keep getting saltier every year - so we can logically conclude this process has not been going on for 3 billion years or so.

Salt and the other chemicals in ocean water also pose another problem for Evolution's 'warm, salty pond' concept - but that is a later Theses.

We are now ready to logically get an idea of how long this process has been occurring. Qualified scientists at ICR find that the maximum age of the oceans is only about 50 million years! This finding DIRECTLY shows that Evolution's billions of years for the age of the oceans is WRONG.

However, you may notice that it also seems to show the Creationist's view of only thousands of years of Earth His-Story wrong as well…

Before throwing your science book and Bible away - consider that we assumed the oceans were pure water, most likely they were not. If the oceans were not pure water - this would SHORTEN the length of time the process has been occurring considerably.

Also consider what we DO know about the Earth. The Earth's crust has great cracks in it at the BOTTOM of the major oceans. These cracks are often UPWARD fissures which indicate that pressure from inside the Earth pushed them up and released something.

What was likely released? It stands to reason that whatever was released should be in the oceans today and should still be found in undisturbed sections of the crust.

The Gizmodo site admits that if all the salt were removed from today's oceans and deposited on land that a layer over 100 feet thick would result. Hmmmm…. maybe we should look for salt layers hundreds of feet thick in the crust of the Earth in areas not cracked… Do we find such layers? YES! In many places around the Earth there are these massive salt deposits.

We have also discovered vast volumes of subterranean water deep in the crust along with the layers of pure salt. We may logically postulate that SOME of that water migrated through the salt layers on its way to mix with any ocean water already on the surface of the earth.

Is there ANY account of some 'breaking up of the great deep' that released massive amounts of water in any book or Book on Earth?

Why yes there is! The Bible's Genesis account of the Flood of Noah contains just such an account witnessed by 8 human beings and recorded for us as a historical fact.

While we cannot PROVE these two factors (plus the higher runoff of water from the land after the Flood) actually do account for most of today's salt concentration in the oceans - we CAN see a path to just thousands of years for this chemical concentration process.

At the same time, we cannot imagine any process that could have removed such a large quantity of salt happening for over 2.9 billion years to make this indicator of a young Earth reach to anywhere near the billions of years Evolutionists claim for the age of the oceans.

Theses Nine:
The Law of Logic and Reason

Premise: 'If - Then' statements can help us determine if a claimed explanation is possible or not by demonstrating the implications of a claim.

We would love to trust that whatever people say is actually True. However, the sage advice of 'Trust but Verify' has always served man well. This applies to me as well! The last thing I desire is for you to simply take what is said in this book at face value alone! I want you to go on from reading this book to doing your own research into this topic!

When the Evolutionist is cornered and questioned about his many 'facts' - which are obviously only unproved and nearly always unprovable claims, many Evolutionists appeal to majority opinion of the scientific community or the 'thousands of papers' written in support of Evolution - submitted to scientific sounding organizations (all run by Evolutionists for Evolutionists). Basically they ask you to 'trust us, because we are the experts'.

Sorry, but I expect thinking people to use their heads for more than growing hair. I want the readers of this book to research what I say and not just swallow it whole. Therefore, I will do the same to the many claims of the Evolutionist.

If what the Evolutionist / Atheist says is a proved 'fact' really is a proved fact - then that statement will stand up to intense scrutiny. We may find some minor flaws in the statement that have to be corrected, for no man is perfect in understanding - but the basic premise should withstand such an examination and survive.

Here is what I mean: I have heard many Evolutionist say that molecules-to-man Evolution is as proved as the Law of Gravity. One Evolutionist said this in his testimony before the Arizona State Board of Education!

Let us put his statement into an 'if-then' format:

If Evolution really is as proved as Gravity, then I will be able to demonstrate that dead chemicals come to life and amoebas change into some other type of critter - because I can demonstrate Gravity in a laboratory and directly observe its effects.

Neither Abiogenesis nor Macro-Evolution can be shown to take place in the laboratory! This man's claim is provably FALSE. He LIED to the School Board - or is so badly mistaken that he should not have been testifying as an expert witness.

Let that sink in:

Gravity can be DEMONSTRATED by experiments that yield consistent results all over the world. The basic premise NEVER fails to yield the expected result.

Evolution of the sort this man was speaking of has NEVER ONCE been shown to happen - NOT EVER!!! Indeed, as shown by an earlier Theses in this book - the Law of Biogenesis PRECLUDES the possibility of Abiogenesis being True!

Using logic, we have PROVED that molecules-to-man is not 'as proved as Gravity'. By a previous Theses we PROVED that what he said was not even possible. This man made a provably false statement to the Board of Education - but no one pointed out the man's false statement to the Board.

When it was my turn to testify; I demonstrated to the Board that a glass jar broken before their eyes would NEVER reassemble itself no matter how much time it was given to do so. I asked them if it then made any sense that something alive could assemble itself. None answered my question and some rolled their eyes or looked down at their papers instead of listening. They had already made up their minds what they were going to do and this time of testimony was merely put on so they could say they considered all views.

The Board went on to authorize teaching Evolution as fact to the kids of Arizona. Why? Because the majority of those testifying that day supported Evolution and had alphabet soup after their names - NOT because what they spouted was logically True.

This is the reason for this book:

To arm enough people with the Truth so that they can challenge the 'experts' and show school boards the complete fallacy that is now taught as fact to kids across the United States and sadly across the world.

While we wait for Truth to ultimately win - people who learn the Truth can also teach their own kids so they are at least aware of the lies they will be told in public schools

Sadly, even if you do show Evolution impossible - most school boards will not support you - but choose to bow to majority pressure.

Here is an idea - why don't YOU run for your local school board? Believe me, after arguing with them for years - they are not the sharpest tacks in the box nor are they paragons of virtue. They are politicians, many may have started out with nobel purposes - but the ones I have tangled with have strayed very far from that path.

All you need is a list of registered voters and the gumption to go talk to them and educate them on why Evolution is wrong. The issue that will likely get you elected is showing how wasteful school boards are with tax money.

In Aiken, SC - the County Superintendent of Schools was paid more than the Governor of the whole State! She had an assistant that did all her work - and he had nine underlings to do all his work - and THEY each had an assistant to do all their work!!

LESS THAN 65% of each tax dollar went to support teachers actually teaching in the classroom! At least 35% was burned up in overhead costs!

The Attorney General of South Carolina said publicly that ANY charity that does not use at least 80% of the funds they collect for the stated purpose is a SCAM charity. So-called 'public education' is a cash cow for Administrators and school boards are largely a steering committee for fat contracts for themselves and their friends.

True education is one of the last things on their minds.

Theses Ten:
The Law / Principle of Uniformity of Process as Applied to Layer Deposition

Premise: An orderly process undisturbed will yield predictable results and logical conclusions may be drawn from those results even if the process itself has not been observed. We will apply this concept to the sediment layer deposition by water issue in this theses.

The concept is a bit complicated, but here is a word-picture you may be able to relate to. Let us imagine that a child changes his stinky socks every day and drops them onto a pile at the end of his bed every day - we can logically conclude that we will generally find the socks dropped on the first day at the bottom of the pile.

Using the same logic, we can conclude that; if over an unknown period of time there has been a deposition of layers of dirt, sand, rock, etc. - and these layers have NOT been disturbed: The OLDEST layer will be found on the bottom of the pile.

Further, it stands to reason that ANY objects found in such a bottom or lower layer were deposited there at the same time and by the same process as the layer itself.

Evolutionists AGREE that this is a trustworthy principle or even a Law. It is foundational to their 'Geologic Column' claims.

At any rate, let us look at the two diametrically opposed accounts regarding how the observed layers we see all over the Earth came to be the way they are found.

After we look at the general concept - a Theses in itself - we will look into the topic of Out Of Place ARTifacts or OOPARTs. Each valid OOPART is a Theses on its own, each capable of showing that the timescale of the Evolutionist is not to be trusted - and if there is not enough time for their made up story to have happened: It did not happen.

The Earth is literally covered with layers of many different types of materials. Layers of salt, coal, limestone, hardened lava and many more are found. Some layers span entire continents and even stretch across oceans to be found on other continents.

Some layers are so deep that people may find it hard to believe that they could have been deposited quickly. Does the composition of the layers themselves tell us anything regarding how fast the layer was laid down?

Many layers - be they coal, limestone, chalk or salt exhibit the very unusual characteristic of being almost pure substance of that layer! How could this be if the layers very slowly accumulated over millions of years? Surely the primary substance would be thoroughly contaminated with many other things that wound up in the layer. Even the Evolutionists scratch their scalps in amazement at this phenomena - but offer NO plausible explanation how this could have happened.

However, let us consider a world-wide Flood. ALL these layers that span continents and can be many hundreds of feet thick would have been washed into place very quickly under the same conditions. It is much easier to understand thick layers of nearly pure substance being deposited under these conditions. A world-wide Flood would nicely explain the ACTUAL data we can observe.

Do we have ANY historical or His-Storical accounts of such a Flood? Why yes we do! We have BOTH! There are some 300 'great flood' accounts from every inhabitable area on Earth. ALL fall into the category of legend EXCEPT for one - which explains in detail HOW the Flood occurred. Those details ALSO fit nicely with many other 'hard to figure out' features of the Earth - such as the cracks at the bottom of the oceans and the not-in-equilibrium salt concentration of the oceans.

The Bible's Genesis chapters 6-8 give us these details. Not only is the HOW explained, but the WHY for the Flood is explained as well. In this case there are HUMAN eye-witnesses that survive the Flood and bring us the record.

It is notable that only Genesis 1:1 through 1:24 - up until the creation of Adam - do we have to depend on God telling Adam the Truth about the Creation of the Universe, Earth and life on Earth.

ONLY 24 verses in the ENTIRE Bible must be taken SOLELY on FAITH that the unseen Being many call God is telling the Truth. ALL the rest of the Bible - it is possible the HUMAN eye witness is telling us the Truth.

I point this out, not that God is not ultimately Trustworthy - but because so many want to ONLY believe what a human testifies to.

Contrast this with Evolution concepts that claim the Universe is 20 billion years old and the Earth is 5 billion years old. According to this story - and it is just a made up STORY - man did not even appear until 2-3 million years ago and could not reason well enough to discern what had gone on before until Charles Darwin came along about 150 years ago.

This means we have to TRUST that man can look at rock layers he did not see laid down and somehow tell us without error exactly what took place during all those billions of years.

Excuse me if I pause here to LAUGH OUT LOUD at anyone who tells me this with a straight face!

More than 99.9999% of THEIR STORY we must take on FAITH alone that they have interpreted the evidence correctly.

In contrast, 99.9999% of the Bible's His-Story involves human EYE-WITNESSES and their statements which have NEVER BEEN PROVED WRONG!

Your choice - but I would say YOU are being far more religious if you choose to believe the Atheist / Evolutionist.

Theses Eleven through Twenty: Examples of Out Of Place ARTifacts a.k.a. OOPARTs

Premise: If an OOPART is found in a layer of rock supposedly tens of thousands or hundreds of millions or billions of years old - and in that undisturbed layer - an obviously human made object is found - then it is logical to believe that the object was deposited by the same means that resulted in the layer being deposited.

If the OOPART is of obvious human manufacture - it logically follows that the layer (and all those undisturbed layers above it) cannot be many millions or billions of years old and Evolution's timeline is proved WRONG.

It further proves that Evolution as a governing theory is also dead WRONG.

If Evolution is wrong - then logically another governing theory must be RIGHT.

If there is only one other governing theory available that is NOT a mix of Evolution and Creation.... then that theory would be right and - like it or not - Young Earth Creation is that theory.

The mere existence of these objects in the layer is IS enough to establish it's recent deposition. Here is why. My family a while back went 'Gem Mining' in the outskirts of Asheville, TN. This entails either buying a bucket of freshly mined dirt/rocks and hope for the best - or - buying a bucket or tub of dirt with gem rocks already put there by the human owners.

In both cases I could take a close-up picture of just the dirt / rocks leaving out the human made bucket they were in. I could then spin a story regarding how long those rocks had been there and you would only be able to judge my claim on whether or not you believed me. If I then expanded the shot to include not only the bucket, but the mining business building and the cars in the parking lot - then told you the rocks had been there millions of years - you would rightly laugh and walk away.

In the 'staged find' buckets - there was a chance you would find a man-made coin redeemable for a prize. One bucket in ten had the coin hidden in it, read the promotion. If I were to find an obviously man-made coin in the bucket - again I would KNOW that the coin had been recently deposited.

You may say that the above example is not one in nature. Ok. In Hawaii they have lava flows that sometimes roll over man-made structures. Many times the item is burned, but if it is metal - oftentimes part survives and becomes an OOPART. One could film an entire documentary stating that this lava flow is millions of years old and show ONLY the parts where no human made objects appear. The sight of the roof of an old tin shed INSTANTLY would disprove this claim.

If you were to confront the person who made the previous claim with such evidence - they would either have to:

1) Admit they were wrong
2) Examine your find and explain it away
3) Ignore your evidence and stick to their story

Let's see what the Evolutionists decide to do - and it is NOT #1.

You see, logic is a cold hearted killer of incorrect theories. If any single one of the following objects were actually found where they were said to be found - it does not matter if the side of the argument decimated by the find agrees to the significance or not; the debate is OVER since their timeline is WRONG and there is not enough time for maggots to turn into mountain lions.

These artifacts are generally completely ignored or openly mocked by Evolutionists because each one demolishes their fairy tale's timeline and they know it. They also know that they have no other answer for a genuine OOPART.

This cannot be stressed enough! This is why I have said it four times in four paragraphs:

If ANY of the following OOPARTs is genuine: Evolution's timescale balloon of billions of years is popped and completely discredited.

This is why you will NOT find the topic even mentioned in the standard public school 'science' textbook - because no challenge to the Evolution Religion is tolerated.

Any who do attempt to take on the Evolutionist's religious construct of the unobserved past will find themselves smeared, ignored, mocked and opposed by the very ones who have taken a solemn oath - often with their hand on the Bible or in God's Name - to provide our children with valid learning materials.

I fought the Aiken County South Carolina school board for five years. That fight somehow garnered national attention. None other than ABC News anchor Peter Jennings featured the fight in Chapter One of what turned out to be his final book: In Search of America.

The book had many chapters and six of them were televised including mine. They shot more than 60 hours of film, yet I was actually given only about 5 minutes of the 42 minutes of content. To be fair, I was given more space and a more balanced treatment in the book. In the movie I was portrayed as a concerned and dedicated parent (true), but either uninformed (false) or unwilling to acknowledge the 'overwhelming evidence' proving Evolution (also false - because there IS none).

The one thing the ABC researchers putting the book / film together could not get over was the fact that - to quote one of them, "You are so smart! How can you not believe in Evolution?"

I replied, "I disbelieve Evolution BECAUSE I am so smart!"

I actually LOOK at ALL the evidence and see if it actually aligns with real science - and it DOES NOT.

I have looked into these things for very many years and have found NOT EVEN ONE thing claimed by Evolutionists to be either right or even possible. All are in violation of some Natural Law(s) or have an alternate explanation that fits the actual evidence far better.

On with the OOPARTs!

The examples I use come from various internet sites and information I have run across in more than 30 years of looking into this topic. I try to use websites NOT friendly to Creationists or indifferent to the Creation-Evolution debate so that critics cannot claim that I only use Creationist biased sites.

Warning!

SOME of the websites claim that aliens brought life to earth! I used their websites ONLY for the data - NOT because I believe anything else they may say on the website. I said earlier that OOPARTs are completely ignored and suppressed by Evolutionists, so I had to go to some other sites that I would not otherwise go to.

I use their information under the 'Fair Use' guideline - and in most cases it was taken by them from another source such as an old newspaper article. Do a search yourself! Just type in OOPARTs into your search bar.

Theses Twelve:
Nampa Clay Figurine Found in 1889

Details: Men drilling a water well discovered in the waste dirt / debris a one inch high clay figure of a human.

According to Evolutionists now days: The layer in which it was found is claimed to be about 2 million years old. According to their timeline, man had barely learned to use stone tools - 'art' or toys would come much later.

How is it that such a figure from an obviously UNDISTURBED layer would contain this Out Of Place ARTifact?

These men were not looking for anything in particular, they were drilling a well.

Darwin's book was still largely unknown in America in 1889 and there was no such thing as a 'Creationist' then because there was no need for one.

Virtually EVERYONE at that time believed the straight-forward account of the Bible - and so were Creationists by default.

People had to have their common sense trained out of them over several generations for Evolution to get and hold the mental and spiritual territory it has today.

We MUST retake this territory. Evolutionists will not go away without a scorched earth fight. We will have to simply beat them with the Truth, logic and firm reliance on the actual facts properly interpreted.

Theses Thirteen: Screws and Threaded Bolts Found Embedded in Rocks

Details: In 1996 and again in 2009; screws were found embedded in rock supposedly hundreds of millions of years old.

Both of these finds were in Russia, but I - myself - have found a threaded bolt embedded in rock in the small town of Channahon, Illinois when living there from 2005-2008.

I also found very small dinosaur footprints in the same layer as the screw, though the rock composition was different - so I do not claim the footprints and the screw in the rock are from the same time period.

My 'rock containing a threaded bolt' OOPART was near a bridge and so the rock could easily have been part of the foundational materials for the bridge brought in from a quarry.

The point cannot be made too many times - if a MAN-MADE object is found in layers claimed to be tens of thousands or hundreds of millions or even billions of years old - the Evolution timescale is completely invalidated.

The only other possibility is ALIENS from another planet! The website messagetoeagle.com DOES offer this as a possible explanation -

because it does NOT want to say the layers these OOPARTs are found in cannot be so old.

This website promotes the idea of aliens coming to Earth. I do NOT believe this happened, I just want to use sites that are not biased towards Young Earth Creation so my critics cannot claim they are the only sources I use.

The BEST explanation is that MEN MADE these objects and they were DEPOSITED where they were found when a great Flood washed everything away and created the very deep layers of sedimentary rock, coal, etc. all over the course of a year or so - about 4000 years ago.

The eyewitness accounts of a man exist to tell us this event happened in a Book never proved wrong in thousands of years of active effort to prove it wrong.

Why not believe an account NEVER proved wrong?

Theses Fourteen:
Metal Bell With Inlaid Silver Design Found in Solid Rock Layer

Details: In 1851 in Dorchester - fully EIGHT years BEFORE Darwin published his infamous book - this OOPART was found in a fifteen foot thick layer of solid sedimentary rock.

The article was featured in…. wait for it….. Scientific American?!

The article even mentions Tubal-Cain; the man who was an 'artificer in brass' in the first few Chapters of a Book… what was that Book again? Oh, yes!

The BIBLE!

Apparently real, accredited scientists in the 1850s could still be YOUNG EARTH Bible believing Christians.

Why not today?

The article DOES NOT mention any age for this layer, because the theory of evolution existed only in the fertile minds of a few Atheists in Europe then. Darwin had gotten back from his voyage and wrote his book, yet was so worried of it's consequences twelve years later that he had not yet published and would not publish for another 8 years.

Evolutionists would LATER give a date to this layer. How old would these FALLIBLE MEN say the layer was - even though this OOPART existed and was apparently known by the science community? They would say this layer was over 500 million years old.

According to the website, this OOPART has recently been studied by the Museum of Fine Arts in Boston - which has access to the laboratory at the Massachusetts Institute of Technology (M.I.T.).

Why assign great age to this layer when they KNEW or should have known there was NO POSSIBLE WAY the layer was that old? It is because Evolution MUST have vast lengths of time available to make minnows turn into monarch butterflies and mushrooms and men. It is as simple as that.

This is the same case as if these snake oil salesmen of Evolution told us a lava flow is millions of years old WHILE standing on a metal shed destroyed by that very lava flow!

Theses Fifteen:
The London, Texas Hammer Fully Encased In Stone

Details: A wooden handle sticking out of a chunk of stone caught the attention of a man in 1934 - just nine years after the famous Scopes 'Monkey Trial' that paved the way for Evolution to take over the public education system.

He broke open the rock to find the handle was attached to a metal hammerhead! The metal has now been sampled and analyzed. It is a very sophisticated alloy not previously known to man. It is a form of stainless steel, but **NOT** the stainless steel men make today - in some ways it is **FAR SUPERIOR**.

Though the rock the hammer was inside of was a loose chunk from the surrounding rock - the rock WAS consistent with the rocks from that layer and found in close proximity of that layer.

The layer had been previously dated by the Priests of the Evolution Religion as being 100 million years old - which also means the dinosaurs were 'ruling the world' right about then.

Take your pick of possibilities; either:

1) Aliens left the hammer there and the rock formed around it
2) Dinosaurs made the hammer and the rock formed around it

3) Man made the hammer and Noah's Flood formed the rock around it

4) The hammer made itself and the rock formed around it

Folks, once again - the BEST fit for this OOPART being where it was and in the condition it was in, is #3.

The man who now has this hammer in his collection has also found a fossilized human finger in the same area…. and a dinosaur footprint over the top of a human footprint! This find will be covered in Theses 17.

Theses Sixteen:
The Ica Stones of Peru

Details: Some 50,000 carved stones are now known to exist from various sites in Peru. MANY of these stones show pictures of dinosaurs and more interestingly: MEN interacting with dinosaurs!

Men have now made pictures of dinosaurs for almost 170 years from bones and reconstructed skeletons. When you compare these 'modern' pictures with those found on the Ica Stones; they are virtually the same! How did these people who lived in this area within the last 5000 years know what dinosaurs were supposed to look like when they did not have the skeletons to work with?

The stones tell us the answer: These people SAW these creatures AND domesticated them AND rode them AND killed (and possibly ate) them.

Of course, this flies in the face of the Evolutionist's tall tale - so they simply dismiss yet another trove of OOPARTs out of hand; without so much as actually examining them! They call them ALL fakes and those who consider them as valid evidence 'fringe types'. Name calling - even if the one being called the name deserves the name called - DOES NOT automatically mean the OOPART is fake!

Remember! God says in His Word that He would use the foolishness of this world to shame the wise. I will grant you that SOME of those pushing various OOPARTs are out of the main stream - many consider me a 'fringe type'. However, ALL of us are just imperfect messengers - it is the MESSAGE that needs to be examined; NOT the messenger. That message clearly indicates man and dinosaur co-existed - at least in this area.

Isn't it amazing that ALL evidence that throws the least bit of doubt on the Religion of Evolution is automatically labeled fraudulent without so much as a cursory examination? One site did investigate the Ica Stones…. sort of. What they did was to show how fraudulent Ica Stones - which do exist for sale to tourists - could be made.

They concluded (wrongly) that since a person COULD make a fake Ica Stone that ALL of the stones found BEFORE there was a market for them are fake - without ever actually looking at a REAL Ica Stone. Who died and left them in charge of determining all things for all people for all time?

Dr. Dennis Swift has written a detailed book, Secrets of the Ica Stones that shows the difference between a real one and one that has been faked. The differences are real and they are easy to detect by an expert.

Why are the claims of skeptics NOT open to examination, yet everyone else's claims and evidence can be mocked or ignored - if their evidence disagrees with Evolution in the slightest degree?

It is time to THINK FOR YOURSELF folks!

Theses Seventeen:
Dinosaur Steps On A Man's Footprint!

Details: The entire topic of human footprints being found amid dinosaurs - and the recent finding of a dinosaur stepping on a man's footprint - underscores the importance of OOPARTs.

Famous Evolutionists have said that if it could be truly proven that men existed with dinosaurs, the Theory (Religion) of Evolution would be at least seriously undermined (some say destroyed).

Consequently, the critics of these finds expend unlimited resources to discredit every single find - be it obviously human fossil footprints along side of or in the same layer as dinosaur prints or the find showing a dinosaur stepping on a man's footprint.

Their favorite tactic is to discredit the messenger and turn focus from the find itself (the message). Should this book become widely read - be sure every odd detail of my life will be trotted out and amplified 1000 times to try to get you to take your eyes off the message and laugh at the messenger. This is why ABC's In Search of America episode about me starts out in a good old fashioned tent revival meeting. They want to make the case - before ANY other evidence is introduced - that my challenge to Evolution was religious and not scientific. This tactic is called 'poisoning the well' - saying to their audience that this guy is a religious nut, probably sincere, but a nut just the same; but we will cover him anyway for a laugh or two….

I have nothing against tent revivals! They are one of the many ways people come to know the Lord. I personally have never attended one - so why would ABC deem it fair to START the show with fully 3 minutes of coverage? It was because I acknowledged to them that I was a fundamentalist Christian - so THEY made the jump to tent revivals, etc. I was surprised that they did not trot out rattle snakes since I was attending a Pentecostal Church at the time.

As usual, it comes down to who you choose to believe - those who say these finds are genuine or those whose entire RELIGIOUS LIFE is based on Evolution being true who say EVERY SINGLE FIND is a hoax or misinterpretation; except those they can contort into some kind of supporting material for Evolution.

Remember: When the Leakey family found human tracks in fossilized volcanic ash many decades ago - NO REAL scrutiny was given to the find! This is because they were part of the club and so whatever they claimed was given instant acceptance and credibility. These prints have now been reburied - to protect them, of course….. or to keep others who may disagree with Evolution from examining them….. maybe they found a tin roof nearby….

A fossil of a dinosaur stepping on a man's footprint has been found.

A cast of that fossil may be looked at in Glen Rose, Texas at a museum - along with more casts of fossil human footprints and actual fossils.

At the behest of those wishing to discredit the find, the New York Times (NYTs) proclaimed the find suspect. As per usual, they based their suspicion on the MESSENGER and NOT on the find itself.

According to livingdinos.com - a series of 6 human footprints in a left-right-left sequence are found in a rock slab with clearly dinosaur footprints nearby.

The location is in Turkmenistan - the former Soviet Union. Apparently even the ATHEIST government that controlled the media when this was found mused that perhaps men mingled with dinosaurs!

A quick internet search indicates sites in Africa, the United Kingdom and several other areas in North America have their own human footprint fossil in dinosaur layer finds.

This begs the question: Are ALL of them fake? The NYTs would have you believe so. It appears that every time ANY find that challenges Evolution in any way is brought to light - the NYTs is quick to tell us it has been debunked. Have they EVER even left New York City to come to their conclusions? No. All they need to know is that it somehow discredits their gospel message of billions of years of Earth age....

These people are JOURNALISTS - just exactly what makes them the experts on fossils or the judges of these finds? WHY do they get to interview ONLY those with a vested interest in protecting Evolution and then arbitrarily decide a find is fake?

You simply MUST stop letting them think for you!

After all that Wikileaks has revealed about the major media in general and the NYTs in particular from the 2016 presidential campaign - I think it is the New York Times that should be SUSPECT until THEY are proved right!

Look up the pictures online yourself or go to Glenrose, TX!

Read about it from friendly and unfriendly sites.

Make up your own mind!

Bottom line - we DO find obviously human tracks in the same layers as dinosaurs not just in Glenrose, but all over the world.

Are ALL fakes or misinterpretations? - or are the Evolutionist / Atheists just circling the wagons trying to save their Religion?

Theses Eighteen:
Brass Bell Found In Lump Of Coal

Details: In 1944, a ten year old boy dropped a lump of coal in his basement and it broke open revealing a brass bell complete with iron clapper. The brass is of a unique alloy and shows a remarkable knowledge of the use of alloying metals with other substances.

The seam of coal from which the coal was mined had by this time been 'confidently dated' by 'professional geologists' to be about 300 million years old. Though generally ignored or 'explained away' by 'professional scientists' - this article on humansarefree.com claims that there are likely thousands of OOPARTs squirreled away gathering dust in small museums or larger museum basements or people's private collections.

I personally have found at least three hard to explain objects in various types of rock without even really searching hard or going to remote places.

The bottom line is that ANY of these OOPARTs - if actually proved to be True, i.e. found as they are described to have been found - show the 'millions of years' claim IS only a fairytale.

NONE of them has been proved to be fake or a false claim. SOME of them have had articles attempting to cast doubt on the OOPART. MOST are simply ignored completely as if the Evolution community sheriff standing in a puddle of blood at a crime scene is saying, "Nothing to see here! Move along!"

There IS something to see…. and what has taken place IS a crime…. the Truth is getting murdered every day Evolution is allowed to sit on Truth's throne.

Theses Nineteen:
Ten Inch Gold Chain Found In Another Lump Of Coal

Details: Mrs. Culp was breaking coal into smaller pieces to burn when she cracked one lump open and found a ten inch long GOLD chain. The gold was analyzed and found to be 8 carat, meaning once again that the human that made the chain understood how to alloy metals.

Gold for jewelry is generally alloyed to make it stronger for delicate design. She described the design of the chain as being of "quaint workmanship" in the June 11, 1891 Morrisonville (Illinois) Times article.

The article had no information regarding the Evolutionists estimated age of the coal - because the Evolution Religion was still in its infancy in 1891 in this country - but we can be sure it would be claimed to be hundreds of millions of years old today.

Here is yet another OOPART that defies the evolution timeline. Who are you going to believe: A woman with NO REASON TO LIE about a strange find BEFORE there were any Evolution Religionists to anger - or so-called 'experts' who 'know for a fact' that a woman they NEVER MET is lying for a reason SHE COULD NOT HAVE KNOWN.

I might understand an Evolutionist taking me to task over my finds today because I am an unapologetic Young Earth Creationist, but in Mrs. Culp's day there was no such controversy. An Evolutionist might claim that I have a reason to lie about my finds, but what of Mrs. Culp? By the time the Evolution Religion formally replaced real science in public schools in the 1950s, Mrs. Culp had been long dead.

The account reads just like what it is: A local woman found something strange in a lump of coal that any five year old knows should not have been there. A human interest newspaper story with NO AGENDA behind it for EITHER side in the current day debate that did not exist then was published.

There is no valid reason to disbelieve her story - especially when combined with the several other finds of OOPARTs in coal.

Theses Twenty:
My Own OOPART Finds:

Details: I have already mentioned finding strange objects in various rocks. One was the screw or bolt discussed before. Another was several pieces of rusted but obviously 'wire' running through pure limestone. In the same pieces of limestone rubble I found what appears to be a dime sized iron button - but could also be a blob of slag. In either case: What is some kind of iron bearing metal doing in a layer of pure limestone that is free of any other oddities except these rusted pieces?

I also found - just yards from where I found the screw in the rock - a number of fossil footprints of three and four toed baby (or just very small) dinosaurs. They all have sharp claw prints made in a clay-and-sandstone matrix. The best one shows either:

+ a five toed critter and is a single print

or

+ it is a four toed critter that then either stepped on its own first track

or

+ the first track was stepped on by a different critter

In any event, the print is pristine and you can see the toenails clearly. Yet another print is about the size of a quarter and shows a three-toed critter that was apparently running - as the middle toe dragged a bit as it was withdrawn from the mud. This dinosaur would have been about the size of a pigeon.

Still one more is a very clear raptor partial print - the curved claw making a curved hole ending in a sharp point in the mud-now-turned-into-stone piece. One geologist looked at it and said it was a natural phenomenon, but when I pressed him on how a CURVED hole tapering to a sharp point that looks for the world like a claw could be made by natural means, he had no answer.

ALL of these were found in Illinois - the limestone with rusted iron at the place where I worked had been brought from some quarry and the footprints near where I lived in Channahon seemed to have been made in situ.

I work as a training instructor for a nuclear power plant. In my position; lying is punishable by dismissal and barring any further work in that industry.

Yet the accounts I lay out above I post for the world to see and judge. If you care to, you can research my credentials and interview folks to see if I am a serial liar - or if I am one of the more honest ones people have known.

Evolution is a completely wrong interpretation of the available evidence. For it to have any chance of being correct - Evolution's proponents MUST label ALL OOPART finds as frauds and ALL those who give their account as liars or kooks or misinterpreting or misrepresenting their discoveries.

I will gladly take a lie detector test on public television regarding my finds if they want to pay for it.

Theses Twenty One through Twenty Five: The Orbits, Origins and Deaths of Comets and Super Nova Star Deaths

Premise: Comet orbits and destruction as well as star death by super nova show the Evolution timeline is not plausible.

Theses 21: A short period comet is one that - as the name implies - circles our sun in a 'short-period' or 'short timeframe'; usually defined as a 'less than 100 year' cycle. Halley's Comet would be a classic example as it's orbit is between 75-76 years.

Each time any comet makes a pass around the sun; a portion of its total mass is lost. We see this as the famous 'tail' strewn out behind the comet's main core.

There is no known mechanism to replace this lost mass.

It can therefore be logically determined that if a short period comet loses even 0.1% of its mass on each trip around the sun - it can only make a total of 1000 trips before it is COMPLETELY gone. Some astronomers believe a comet may lose up to 5% of its mass each trip for those that travel close to the sun. Such a short term comet may make a maximum of 20-30 orbits before it is gone.

Halley's Comet is still here; having last appeared in 1986 when I was 25 years old and is expected to return the year I turn 100 in 2061. This means Halley's Comet has only been making the trip for a FEW

THOUSAND YEARS since it still has a quite massive core and magnificent tail.

Evolutionists desperately want to believe the Universe is twenty billion years old, but if that were really True: There should be NO short period comets comets left. Even if Halley's Comet only lost 0.1% of its mass each trip - it could only possibly last about 75,000 years before it would be completely gone. That it still exists and is still massive means it is NOT even 75,000 years old!

Theses 22: There are very many short period comets that we know of. As I write this part of this book, my daughter told me there will be one visible for the next few nights. This means the Evolutionists have another problem besides the fact that we still have short term comets: They need a SOURCE for new comets.

Go figure; they have come up with yet another 'theory' to cover-up this major problem: The Kuiper Belt and The Oort Cloud!

Yes, boys and girls - ANOTHER unseen and unprovable source of billions and billions of comets - some of which just happen to (for no reason whatsoever) just fall into a very stable orbit around our sun!

The Kuiper Belt conveniently is supposedly located somewhere out near Uranus and The Oort Cloud supposedly surrounds the entire solar system. Neither of these concepts has been proven - yet they are discussed as if they have been proven in all the Evolution friendly articles.

Those that try to reign in these wild-eyed magicians are labeled 'Deniers!' and summarily kicked out of the club.

Theses 23: The orbits of these heavenly bodies are very stable and often predictable - hence the reason we will know to look for Halley's Comet again in 2061.

IF these comets were originally in some kind of orbit in either The Kuiper Belt or The Oort Cloud - some kind of collision would have had to take place with some other heavenly body to knock said comet off its current course and into the orbit it now follows - or a violation of the Law of Conservation of Momentum covered earlier would exist.

Yet if ALL of the matter in the Universe came from a single 'Big Bang' - ALL of the matter would have flown out from a common center into the nothingness of space. ALL objects would be heading away from that center and NEVER be on a collision course with anything!

In other words, the Evolutionist has no valid mechanism to cause these comets to behave they way they are observed to behave.

Yet I am told that I must believe their every musing with unquestioned fealty. No. I will not. I have a brain and it still works. I will use it to examine ALL the evidence and draw my own conclusions.

Theses 24: The comet Shoemaker-Levy was also a short period comet. It did not die a slow death of gradually losing its mass to light up the night sky.

This comet crashed into Jupiter, instantly ending its existence forever. Here is yet another way to lose comets long before they 'burn away'. Still, we find very many short period comets - why so many in a supposedly ancient Universe? Simple! The Universe is YOUNG!

If you say the Universe is but thousands of years old and the Creator set these objects in their path - ALL issues are nicely solved!

The Evolutionist CANNOT allow for there to be a Creator - so he makes up a derisive term to hurl at his opponents. The Evolutionist claims the Creationist is appealing to God to fill in the gaps.

What the Evolutionist never tells you is that he also fills in the gaps in his theory.... with NOTHING at all!

Both sides have to fill in gaps in knowledge! Take your pick: Put SomeOne in the gaps or put nothing at all in the gaps - that is your real choice.

Theses 25: Star deaths by super-nova! These are truly incredible events that we observe on a relatively frequent basis. Hmmm.... we see star deaths, but we have yet to see even a single star being born outside of Hollyweird..... ummm, I mean Hollywood.

The Evolutionist knows he has a serious problem here, too. What does he decide to do? Why make up another UNPROVED and UNPROVABLE theory of course!

The Atheist astronomer says, "See those huge dust clouds called nebula through our telescopes - well, they are really star nurseries!"

Yes, boys and girls - these clouds of dust condense and... and... and... get really hot.... and.... and... umm.... suddenly ignite and become a star! Oh! - and this has happened for each and every star that exists today - all 100 billion stars in each of the 100 billion galaxies!

Folks - it is time to throw the bull-crap flag on these people!

They suck up many hundreds of millions of YOUR tax dollars to peer into the night sky and all we get for that money is this? Enough! There are better uses for all that money and these people's time.

Does Creation offer a better explanation? That is for you to decide. The formation of stars and origins of comets are things beyond the current capability of man to understand fully. Since that is the case, an appeal to a 'higher power' is not absurd so long as all other views can be shown to be suspect as is the case with the Evolutionist's myth about star birthing.

To some, inserting God so often may not be satisfying or they may call it a cop-out; but the alternative is to believe their even more crazy explanations. At least Creation explanations align with known Natural Law and Logic - even if ultimately we have to say the Creator chose to do it this way.

The Bible simply says Creator God made the great Lights - one (our Sun) to rule the day and one (our moon) to rule the night - and He made the stars also. The word translated into English as stars can include all lesser heavenly bodies like comets and meteors as well.

If man could explain all the things in the Creation, God would not be God - would He? The more man cannot explain - the Greater that God would be.... There is a great deal we mere men have no real clue about! This fact alone testifies to the Greatness of God!

Summary: Evolution's timeline is fatally wounded by a genuine analysis of all aspects of comets and stars.

The Atheist / Evolutionist has only unproved theories that ultimately violate Natural Laws.

Creation's Creator God could do things any way He saw fit - and all the actual evidence indicates He did Create only thousands of years ago.

Theses Twenty Six through Thirty: The Earth's Moon

Premise: If Earth's moon has a logical, naturalistic, God-less source - then Evolution should have a logical explanation for all aspects of our moon. Trouble is - Evolution does not have such a case to make; only more guesses and theories.

Earth's moon stands each night in silent testimony to Creator God. King David told us in the Psalms to look into the night sky and it would teach us. Humility is one thing it teaches us - for the vastness of this Universe and all that is made by One Being! No wonder there are many hymns of the faith that extol God's greatness in Creation.

My father's favorite hymn was; 'How Great Thou Art'. Its opening lines are:

'O Lord my God! When I in awesome wonder
Consider all the worlds Thy Hands have made!
I see the stars! I hear the rolling thunder!
Thy Power throughout the Universe displayed!
Then sings my soul! My Savior God to Thee!
How Great Thou Art! How Great Thou Art!

Perhaps we all need a good long stare at the night sky instead of staring mindlessly at the TV and then going to bed. Truly Lord, How Great Thou Art!!

Theses 26: Moon rocks are of a fundamentally different make up than those of Earth according to many websites. Some sites say there is a general similarity of some of the rocks, but the proportions of the various similar rocks are different. The general conclusion is that the Earth did NOT 'burp' the moon into being.

General similarity is NOT a help to Evolutionists. Remember, a Common Designer answers that charge far better than a common source for which the math and physics and chemistry do not work.

The Big Bang could be a common source, BUT that still leaves the orbit trajectory and dissimilarities of materials as insurmountable issues.... not to mention the Big Bang's cosmic egg that exploded has no cosmic chicken to lay it.

I actually saw a National Geographic for Kids magazine article that admitted they did not know where the moon came from and then ASKED THE KIDS what they thought! The gurus of Evolution Religion could not authoritatively tell the kids where the moon came from and ASKED THEM for help in solving the mystery! - and YOU pay tens of thousands of dollars a year to send your kids to college to be TAUGHT by these people!!

STOP DOING THAT!!

Never once was it even considered by the writers of this article that maybe, just maybe - God made the 'lesser light to rule the night'.

Theses 27: The moon's orbit defies any naturalistic explanation. It could not have been 'captured' while wandering by Earth millions of years ago. The forces that might have tended to capture it would have given it a much different orbit than it has according to known math.

Remember, there is no appreciable friction in outer space - so any capture theories would have to come up with a reason for the current moon orbit versus any capture trajectory that may have ended up with the moon in the orbit it now has.

To do this, astronomers would take the current moon orbit and 're-verse engineer' any possible trajectory that would result in the current orbit. After trying everything they can think of - they have still come up empty!

Ummm... could I give it a try?

"In the beginning God created the heavens and the earth.... and He made the greater light to rule the day and the lesser light to rule the night...."

They scoff and run me out of the room - BUT they have NO explanation at all. At least I have one that MIGHT work. They would rather believe in NOTHING than in God.

Theses 28: The moon's craters still have sharp detail. Even in outer space, gravity would eventually pull these details down if gravity had been acting on these structures for billions of years.

If one lets their public school mind-conditioning go and simply looks at the evidence with an open mind, they would conclude this process of gravity pulling down the craters must not have been happening for very long.

Theses 29: The moon is slowly moving away from the Earth some say as much as an inch or so per year - and will one day head off into outer space.

The moon also is the primary cause of tides on Earth. This means that the moon would have been closer in the past and the tides therefore would have been higher. If the earth really was 5 billion years old and if the moon was in orbit the whole time - it is logical that the moon would have been a LOT closer then than it is today - about 52,000 miles closer.

The moon today is about 239,000 miles away. Gravitational Force would have been exponentially higher between Earth and moon. A closer moon would have caused massive tides in the newly formed oceans and flooded much of the newly formed continents.

Just exactly how does this make sense with Evolution's critters crawling up on land from the ocean, becoming able to breath air, developing legs, etc? These fanciful critters would have drowned twice a day - ending all the 'progress' made by supposed Evolution in an instant.

Theses 30: In 1969, our astronauts were worried they would sink into a projected very deep layer of dust because they were told the moon had been up there for billions of years and our other spacecraft had measured a very dusty solar system in near Earth orbit.

When they touched down, they found only a few inches of dust - i.e. only a few thousands of years worth. Apparently some Creationists got there first and cleaned up the landing site.

Not every Creationist uses this argument - but it sure is strange that we were given two depths of dust; more than a mile deep using the Evolutionist's calculation and a few inches using the Creationist's numbers - and there was no possible way for any man to interfere with the sample until man got to the moon.

The results were NOT ambiguous at all. The depth of dust was exactly in line with a recent Creation event just thousands of years ago - and no where near what the Evolutionist's were predicting.

Summary: If the moon was placed in a specific orbit by a Creator a few thousand years ago; with no need to evolve life on earth from some critter that got oxygen from the water via gills - ALL the problems Evolutionists have disappear for the Creationist. The only trouble for Atheistic Evolutionists is: A Creator God is needed - which rains on their parade in a big way.

Theses Thirty One through Thirty Seven: Other Features from the Universe Showing Recent Creation

Premise: Everywhere we look in the Universe, we see objects and processes and structures consistent with a Recent Creation which could not exist as we see them if Evolution's billions of years were true.

Theses 31: Some have complained to me that the sharp features of the moon's craters is not valid because of the absolute zero of outer space would preclude them being pulled down. Again we see their objection is but a mere claim in the face of the known fact that gravity pulls everything down towards the local center of gravity.

Ok - leaving moon craters behind as one 'agree to disagree' point - then what about the highest mountain on Venus named Maat Mons? It still shows a very detailed crater AND the surface temperature of Venus is over 800 F; hot enough to melt lead.

It seems God has let man discover physical aspects of our existence to prove wrong EVERY carping complaint of Evolutionists.

Theses 32: The meteor belt consists of billions of rocky objects we are told.

The fairy tale regarding how Earth, Mercury, Venus and Mars came to be rocky planets - amid the much larger gas giants Jupiter, Saturn, Uranus and Neptune - that fairy tale claims without a shred of proof that a bunch of meteors caught up with each other and melted together. Of course, none of this was witnessed - but Evolutionists insist it is a fact anyway.

Riddle me this, Batman: Why are we not seeing this phenomenon occurring today? Why have not several more planets formed from said meteor belt? Even if this did happen - where did all the water come from that we have here on Earth and NOT on the other rocky planets? Why is Earth so blessed with mega quadrillion gallons of water when the other three rocky planets - on both sides of Earth - got left out when our solar system formed?

All the Evolution Religion has are a series of fabricated 'just-so' stories that can only be believed or not believed.

Theses 33: Our telescopes show us many galaxies that have spirals visible - and some that have not completed even a single full rotation! We have measured how long it takes for many of these galaxies to make a full rotation and found it to be on the order of ummm…. thousands of years…..

Question: If these galaxies ARE KNOWN to rotate fully in just thousands of years and the Universe is supposedly billions of years old - why are we still able to find galaxy spirals at all - and worse, any galaxy that has not even rotated fully one time?

The most logical answer is that the Universe simply is NOT billions of years old, but only thousands.

If only thousands of years old, there is NO TIME for stars to be birthed by some condensing of dust clouds. The WHOLE storyline for Evolution comes apart at the seams.

Theses 34: Jupiter's moons are very odd. One named Io is volcanic while its neighbor Europa is an ice ball. They do not all orbit in the same plane or direction, are not of the same materials, etc. How is it that the conditions that formed our Solar System - according to Evolutionists - formed rocky moons around a gas giant planet?

Theses 35: Once again it is as if a gracious Creator God that WANTS all men - even Athiest / Evolutionist astronomers - to understand that He IS the Creator: He put ALL this evidence right in front of their eyes! This is also why they WILL be held accountable for their unbelief even if there was no Bible or Creationists to witness to them (Romans 1).

This is why the Bible says they are without excuse for their unbelief in a Creator God. He still sends them people to witness to them and still makes His Church wait for His Return and keep churches open for them to come hear the Truth. This is part of why this book is being written! - as yet another testimony to them!! What a GRACIOUS God! People spurn Him at their own peril.

Theses 36: The limits of man's ability to actually measure distances to the stars is only about 400 light years according to the YouTube video channel Brain Stuff - NOT a Creation friendly site. After that, man ESTIMATES using brightness of the stars.

Excuse me? I can put two cars the same distance away from the observer - one with its lights on bright and the other with its lights on low beam. According to this method, it would be determined that the car with its high beams on would be closer when in fact they are the same distance away.

A man (me) with NO astronomy degree - or any degree for that matter - is able to refute their claim in 30 seconds? The Evolutionist's answer? - well, "How dare I challenge them?! WE graduated from Harvard and Princeton! We will not deign to give an answer to you - for obviously you would not understand our answer anyway!" This IS the only answer they give - none!

When Jesus was questioned by the Wise Guys of His day, asking Him by what Authority He did what He was doing; miracles - since they could not deny they WERE miracles, He said, "I will ask you a question: John's baptism - was it from Heaven or was it from men?"

They debated amongst themselves - knowing either answer would result in humiliation before the crowds - so they said, "We cannot tell." - i.e. we don't know.

Jesus answered them - as portrayed in the movie 'Jesus of Nazareth', "You tell me nothing….. neither will I tell you by what Authority I do these things."

Atheist astronomers DESIRING to cast doubt on the Bible as His-Story tell us they KNOW FOR A FACT that these stars are millions of light years away - when they DO NOT KNOW. They then say that if it has taken that starlight so long to get to us, then the Bible's timeline is wrong - therefore there is no God - for you Christians say that 'God' would not, indeed cannot lie - yet He did lie in telling us the Earth is young.

The old saying goes, "You can't fix stupid." - and after a while arguing with these folks, you come to realize that the problem is NOT that they do not know they are wrong - they simply cannot bring themselves to admit it and accept the alternative.

With regard to starlight somehow proving how far the stars are away from Earth: The fact is they DO NOT know how far the stars are from us. Period.

The maximum distance we can measure by parallax and using trigonometry is about 400 light years - as ADMITTED by an Evolution friendly website.

Therefore, starlight CANNOT be relied on as some kind of 'disproof' of the Bible. If we add in ALL the rest of the evidences for a young Universe, it quickly becomes apparent which side has the better case - that would be the case for a Creator God.

Theses 37: Our sun burns an incredible amount of fuel every second! How much? According to many websites - none of them friendly to Creationist thought - the sun burns several millions of tons of hydrogen a second! One site claims this is not a worry that the sun will run out of fuel. Due to its massive size, the site claims the sun could burn at this rate for 10 billion more years.

Hold on a minute… a couple of points:

1) These calculations ASSUME they KNOW the mechanism of how the sun operates to be nuclear fusion of hydrogen into helium. We do NOT know this to be True. Another mechanism that could explain how the sun operates is gravitational collapse - the very theory of how the sun was born in the first place! - or it could be a combination of factors - or something we humans do not yet understand. If fusion theory is correct, theory tells us there should be a huge number of neutrinos being 'shot out' from the sun. An experiment to detect the neutrinos conducted in the 1960s showed two-thirds TOO FEW neutrinos than our calculations predicted. Other experiments since claim to have corrected this issue - however, we are largely forced to believe the interpretations of 'scientists' whose SOLE MEANS OF INCOME are research grants…. if they do not find SOMETHING that the one paying for the research WANTS them to find - the money dries up and they have to get a real job….

2) Just because the sun might be calculated to be able to last another 10 billion years from today DOES NOT mean it has been burning for billions of years BEFORE today. Creator God would know that a stable power source would be needed for what He intended to do on Earth and simply MADE such a power source. When He was done with it - He could just throw it away or whatever He wanted to do with it.

3) If the sun has been burning for billions of years - a LOT of fuel has been consumed. That fuel would have taken up a LOT of space - i.e. the sun logically would have been MUCH BIGGER in the past. We KNOW that for life on this planet to exist, there has to be an incredibly precise balance of over forty factors. If ANY ONE of these factors were not exactly as it is within its range, life on earth COULD NOT exist. If the sun were much larger and more massive in the supposed billions of years past, the planet Mercury would not exist as it would have been destroyed by the more massive Sun. Conditions on Earth would not have been within the range needed for life to exist - much less evolve!

'Solutions' of Evolutionists cause more problems than they solve.

Theses Thirty Eight:
Law Breaking Sequences of Events for Evolution

Premise: If a Natural Law is True and a proposed theory can be shown to violate that Law, then it is the theory that is wrong and the burden of proof is on those promoting the theory to show that the Law is wrong. Until those pushing a theory dethrone the Law and replace it with their newly elevated theory-now-become-Law, people have an obligation to believe the current Law and its logical implications.

If folks want to entertain their pet theories - they are free to do so PRIVATELY or in their little groups - BUT we must draw the line at the point where they want TAX dollars and public school science class time to push their pet belief system - their RELIGION.

Theses 38: Let us start with the 'beginning' of God-less Evolution and compare it to KNOWN Natural Laws to see if it is logical to believe in Evolution.

Violation ONE: If Evolution be True - then either something came from nothing (impossible) or everything has always existed (an unprovable belief).

Since evolutionists like majority opinion - if pressed, most Evolutionists I have sparred with reluctantly mumble that they believe in the 'something came from nothing ONE TIME, LONG AGO in a Big

Bang' claim. As we have pointed out; this claim is in direct violation of the Laws of Conservation of Mass and Energy.

Violation TWO: Just to see where they have to go next, let us give them a 'self-creating universe'.

Ok, now we have a LOT of dust, energy, particles flying AWAY from the center of the Big Bang into nothing-ness; leaving what we call the Universe behind.

Let me get a strong hit of LSD…. OK…. what do you want me to swallow next? We need planets in orderly solar systems in stable galaxies that somehow evade the army of Black Holes that vacuum up everything for billions of years….

Wow! - even that hit of LSD was not enough! Got some cocaine? Crack? Whiskey?

Evolutionists propose that the exploded parts of the Big Bang 'condensed' into the structures we see today - no proof that this could even happen is offered, it is just thrown out there. At the planetary scale - at least for Earth - they say a bunch of meteorites 'caught up' with each other and melted together. All this LSD induced musing amounts to tens of millions (if not billions) of individual violations of the known Laws of Motion and Angular Momentum and Gravity.

Violation THREE: Ok, since they are stomping their feet and holding their breath - we will give them their magician-less, magically evolved Earth - complete with the millions of QUADRILLIONS of gallons of water on it that they have no idea where it came from except an unobserved collision with a comet - that would have destroyed the earth instead of giving it all the water it has….

Now they need life, but all they have is water and rocks… so they say it rained on the rocks for hundreds of millions of years and made a chemical soup that got hit by lightening and WALLA: Super-Amoebae arises! Ummm….. this is a DIRECT violation of the Law of Biogenesis which DETHRONED the Evolutionist's claimed Law of Spontaneous Generation more than one hundred years ago.

Violation FOUR: Wow…. ok, they are crying now so we will let them continue their fairy tale. Now the super amoebae needs to morph into other types of single-celled critters, plants like a Venus Fly Trap and a banana - and critters like salamanders and fish and birds and dinosaurs and finally people.

Author Frank Paretti gave a talk titled "What We Believe" - available on CD at Focus on the Family - that summed up this whopper as "…from GOO to YOU by way of the ZOO!"

Each time an organism gets MORE COMPLEX - it MUST gain information. Each of the tens of thousands of random, chance mutations that purportedly account for the most complex organisms alive today is a violation of the Law of Entropy that states that systems tend toward more DISORDER without INTELLIGENT intervention.

Hmmmm…. God is Ultimately Intelligent…. maybe He Created??? Naw! That kind of thought is for religious fanatics! Far more likely that

ALL these violations of Natural Law took place at exactly the right time and in exactly the right way for me to be typing this book....

Conclusion: There is a Book that claims there is a being that not only LIES, but is the FATHER OF LYING! It is one thing to claim something is True when you have no proof it is actually True - especially when at every point in your concocted tale; a known Law or general principle or simple logic is violated.

It is another to to TEACH your unproved claims are true to unsuspecting kids in public school at tax payer expense. It is yet another thing when these kids grow up and establish national governments based on these lies.... governments that then start wars or institute abortion as birth control or condone slavery or consign women to only menial tasks and having babies.

There is a terrible social cost when the Religion of Evolution is instituted as national policy - just ask the people of Adolph Hitler's Germany or Joseph Stalin's Soviet Union or Mao's China or Pol Pot's Cambodia or Castro's Cuba.... or.... or.... Each of these mass murdering maniacs believed in some form of Evolution.

It may chap your hide, but Jesus said it best:

"Satan was a murderer from the beginning. He is the Father of lies and when he lies he speaks his native language!"

If you listen to some of these prophets (oops, I mean professors) of the Evolution Religion (oops, I mean Scientific Theory - though it technically does not meet the definition of Scientific Theory….) - these folks have their lines perfectly memorized and recite them with great passion… they obviously believe what they say - or, like actors, are good at making you think they believe what they say.

Trouble is - as you can see by now - they have been DECEIVED or KNOW they are lying. Yes. EVERY LAST ONE OF THEM has been deceived. They are all WRONG.

Some know they have been deceived, yet lack the courage to admit it and face the ostracism that surely will come. These folks fear men more than they fear God. They need our encouragement to flee the slavery of Evolution-ism and our support in starting a new life.

Is this an outrageous statement: To say that ALL Evolutionists are deceived? Yes. It IS outrageous that this LIE has gotten this far! When the reigning Religion of his day got so far out of control that it was worth risking his life to correct - Martin Luther stood against it. In his day, EVERY LAST ONE of those priests and leaders and teachers supporting what had 'evolved' into the Catholic Church of his day were WRONG!

Many before him had stood at the risk of their lives…. and LOST their lives rather than mouth insincere agreement with the tyrants. It is long past time for REAL Christians to stand tall once again and put an end to this preposterous LIE of God-less Evolution!

The Bible says that in the last days; God will send a great delusion upon ungodly men - so that they will believe a lie.

Evolution IS that LIE we were told would come. Hmmmm.... pretty good prediction for an old irrelevant sheep-herders religion....

Theses Thirty Nine and Forty:
The Laws of Mathematics

Premise: 2 + 2 = 4; always has and always will - this is known as a mathematical truth and there are many such truths that bear on this discussion such as the topics of Probability.

Theses 39: I read somewhere that the mathematical odds of something being possible versus impossible was generally agreed to be:

If it had a single chance out of ten to the fiftieth power - it was possible.

That same article looked at the odds of the SIMPLEST KNOWN form of single celled life's protein chains found in its cell nucleus coming together BY SHEER CHANCE.

It was noted that this bacterium had 100 protein chains; each made up of some 100 complex molecules. The author - who was a PhD level person - stated that for this bacteria's 'necessary for life' material to come together by chance in exactly this way was…. ready: More than one times ten to the 3,200,000th power!

Therefore, he could LOGICALLY CONCLUDE that for the critter he was staring at under the microscope to exist was PROOF that Godless Evolution is IMPOSSIBLE!

Some other conclusions would be that if we ever do find life on another planet: God created it there, too! Why He did not tell us He did is another matter - IF we ever 'make contact'. Evolutionists spend billions of dollars looking for E.T. - money and time and intelligence that could be used to solve real problems for the life that already exists on this planet.

Theses 40: In mathematics, any value can be inverted. This is called the reciprocal of the first value. A reciprocal is a type of opposite of the first value. Creation and Evolution are opposite theories. One is completely correct and the other is completely wrong.

Mixing a little of a wrong view with the completely correct view obviously cannot 'improve' the already completely correct concept - it can only mar it. Therefore, ANY attempt to mix any concept from Evolution with the Bible will only result in a corrupted version of the Bible.

The result of mixing some amount of Evolution with Creation has a name: Theistic Evolution - and can be summed up as some form of 'God did it, but He used Evolution to do it.' It is an attempt to bring the two diametrically opposed sides to some kind of compromise peace. It is actually the view that garners the highest percentage in a survey asking people to choose between: Young Earth Creation (YEC), Theistic Evolution (TE), Pure Darwinian God-less Evolution (PDE).

For many years, the general breakdown of such surveys has been about 35% YEC, 55% TE and 10% PDE. Yet it is the PDE folks that have gotten control of the public education system and forced PDE to be taught exclusively! We HAVE become fat and lazy with regard to our responsibility to teach our children True things and follow True things ourselves!

The Bible states it well, "What has Light to do with Darkness?" There can be no peace between the two views - the TE folks currently camped out in no man's land control the debate!

One side will win an ultimate victory by convincing the TE crowd and the other side will be brought to extinction - except when studied by people in the future to see Evolution as an example of how deceived a majority of people can become.

Who is winning the TE crowd over? The YEC numbers keep growing every time the survey is conducted. The TE numbers steadily fall and the PDE numbers stay about the same. Creation IS winning slowly - but like the conclusion of WWII, needs an atomic bombing to decimate the other side's ability to fight.

THAT is the purpose of writing this book. After anyone reads it - they either will have to admit Creation is the CORRECT view and repent before their Creator - or - they will have to admit they simply want to believe what they want to believe in the face of irrefutable arguments. This book eliminates forever the no man's land position as a tenable position, just as Martin Luther's work forced folks to decide for or against the Roman Church.

Getting back on topic: Since Theses 39 shows that the odds of any type of life coming about by chance is absolutely impossible - what if we take the reciprocal of the number in Theses 39? This is sound reasoning since there are ONLY two possibilities out there - either we were Created or we were not.

It would be stated this way: What are the odds that life was CREATED? The odds we were created is far greater than we can imagine! - on the order of ten to the 3.2 millionth power to one that we WERE created!

There is a Book that claims we were created…. hmmmm…. maybe we should dust it off and read it…… and do what the Creator says…..

Theses Forty One through Forty Six: Proofs on the Nuclear Scale

Premise: Atomic structure and elements - since they are the building blocks of everything - should clearly reflect evidence of a Creator.

We have looked at the far reaches of outer space's galaxies and our own solar system. We have looked at some things observable to the naked eye here on earth. Thus far, EVERY thing we have considered has its best explanation rooted in Creation ideas and principles.

If Creation is True; then the clear evidences we have seen so far will be expected to be observed at the atomic level as well.

They are!

Theses 41 and 42: A theoretical force holds atoms together. That force behaves unlike any other force known to man:

What holds atoms together? Even after 75 years of experimenting with atoms and sub-atomic particles; man STILL does not KNOW for sure how it works!

Man has named the force and described its characteristics - but where it comes from or how it is actually generated still baffles us. Yet, we build nuclear power plants and bombs, etc. - confident that what we do know allows this technology to be safe enough to work with.

I teach Nuclear Theory at a commercial nuclear power plant. The curriculum we teach is approved by the United States Nuclear Regulatory Commission (NRC) - and it differs somewhat from what is taught in high school or colleges.

The NRC approved course identifies that protons are found in the nucleus of an atom and that the 'like-like' or '+, +" charges of these protons exert a very strong repelling force due to the Law of Electrostatic Charges.

The course material indicates that magnetic force does not act at this level to any great degree and the gravitational forces between the protons and protons or protons and neutrons or neutrons and neutrons - the only known force trying to hold the atom together - is forty orders of ten SMALLER than the electrostatic force trying to propel the protons away from each other.

The course also tells us that electrostatic, gravitational and magnetic forces are field forces that decay in strength 'exponentially' - by the square of the distance between them. Example: If the distance is doubled, the gravitational force will be only one-fourth as strong. The effects of these three field forces we can demonstrate in a laboratory.

The atom as we understand it and draw it should NOT exist based solely on the three lab-demonstrable field forces.

Since our experiments DO show that we are correct with our models of atoms; humans THEORIZE there must be yet another force that holds the atoms together - one strong enough to overcome the electrostatic repulsion of the protons.

In the NRC course materials, this force is called the Nuclear Force. In colleges it may be called the Strong and Weak Forces. In either case, this force is only theorized to exist - because it CANNOT be demonstrated in a lab and without it, we cannot make our models for nuclear fission work - nor can we explain atomic structure as our experiments reveal it to be.

The Nuclear Force exists in a spherical shape out to a very short distance then drops off to zero. If you superimpose the exponential decay curve of the Electrostatic Force over the 'straight down drop off' Nuclear Force - we see a problem.

The protons attempting to come together to form the nucleus would NEVER have gotten close enough to each other to 'stick' using the Nuclear Force. Long before they ever got close enough for the attractive Nuclear Force to act, the protons would have repelled each other away from each other by the Electrostatic Force.

How then did the protons overcome the Electrostatic Force that would have kept the protons away from each other?

There is ONE - and ONLY ONE - logical solution to this problem:

A Creator desired to build things using atoms.

This Creator could have made the atoms in one of two ways. First He could have created the protons and neutrons and all the different Forces and then Intelligently pushed the protons close enough together and created a 'nuclear glue' force strong enough so the protons would stick together. The other alternative is that the Creator simply made the atoms directly in their current form - knowing that man one day would see the impossibility of atomic structure existing as our experiments show them to exist WITHOUT a Creator God to make the atoms.

The Bible says that He holds all things together and in Him we live and move and have our being. Sounds pretty much like what man has discovered concerning atomic structure. This is but ONE of very many accurate scientific statements found in the Bible - though the principles are not stated in the dry, boring scientific jargon of today.

Enrico Fermi was a nuclear scientist on par with Einstein. The nuclear plant just south of Detroit bears his name. He is widely recognized as the father of modern nuclear power.

When asked after a lecture on nuclear structure how atoms hold together - Dr. Fermi is purported to have answered: "As far as I can tell… Spirit!" The man revered as the father of nuclear power and Einstein's peer could not explain the atom as we know it without appealing to an unexplainable spiritual concept!

Yet those of the Evolution Religion mock the Creationist; saying we are intellectually inferior or outright stupid. Apparently THEY are smarter than Dr. Fermi? Doubtful. If they are, where are the nuclear facilities that bear THEIR names? They do not exist.

These people ARE legends in their own minds and it is this pride that blinds them to the actual Truth.

Theses 43: Earth has nearly all of the known elements; yet no other place in the known Universe has as many. Most only have a few elements.

Why is this one dust speck in an insignificant solar system in an average galaxy in an apparently limitless Universe so blessed, while the rest of the stars and known heavenly bodies have so few elements?

What if a Creator KNEW we would need all these elements to live - so He MADE THEM for us.

The Evolutionist will say I am again inserting God into the Gap of this conundrum - claiming that once again anytime I cannot explain something I just pawn it off on Creator God.

They ARE CORRECT in part of their charge!

The difference is that as a Creationist, I have examined ALL the evidence as fairly as a human can - knowing my own biases and purposely going out of my way to allow no-God explanations their 'day in court'.

Once both sides have rested their cases, I then simply pick which side makes the most sense. It is NOT my fault that in every single instance for more than thirty years the Evolutionist has failed to make a convincing argument!

You see, their objection to my inserting 'God into the Gap(s)' of my understanding presupposes that their explanation is more plausible.

What is their explanation in the case of nuclear structure? Umm…. oh, yeah… once again; they do NOT HAVE ANY explanation at all! All they can say is that they are certain God is NOT the answer!

Most Creationists were once Evolutionists or Theistic Evolutionists! It was only when we reached our limit of tolerance for explanations that could NEVER possibly happen that we bolted from the Evolutionist camp waving the white flag of surrender and defecting to the only other explanation possible: Creation.

ALL of our choices then become: God did it or the stock Evolutionist answer, "Its here isn't it? It happened, move on and do not worry about how. Give us hundreds of years and billions more tax dollars and we promise we will figure it out….. some day." They say this for many of their failures to explain ANY alternative.

I fully admit that I insert God into the gaps of human understanding - but at least I have something to insert into the gap! The Evolutionist inserts NOTHING at all into the gap or their completely nonsensical, hair-brained musings into their gaps.

Theses 44: There is NO viable 'no-God' method to form the heavier elements.

The formation of the 'heavy elements' is another place the Evolutionist inserts no explanation at all into a gap in their overall 'theory'. In fact, they hope you never bring it up.

The element Technetium has never been found to occur naturally on earth. It seems to be present in some stars according to the radiation signature we have detected for those stars.

Why is this such an issue?

There is ONLY ONE known way to make a heavier element apart from fusion. That way is for an atom with too many neutrons to 'beta minus decay'. This means a neutron spits out an electron and changes into a proton. This process DOES occur in the heavier elements quite frequently. Without this process, a breeder reactor that breeds plutonium could not work.

However, Evolutionists NEED to get to those heavier elements first! Technetium not being found on Earth means a rung in their theoretical ladder is missing! Beta minus decay can ONLY produce a SINGLE proton change at a time. In other words, this process can only 'ascend the ladder to higher elements' one rung at a time.

Bypassing Technetium to get to Uranium, etc. would require TWO rungs to be ascended in ONE step. There is NO process yet identified by man and proven experimentally to exist....

There IS a solution to the problem:

God could simply have created ALL the elements in the Universe and distributed them EXACTLY as He intended right away at Creation - including leaving Technetium not found on Earth. He would know that Technetium was not essential to life AND that one day Atheists would not have ANY explanation aside from acknowledging that He had indeed Created.

Once again, the Creationist LOGICALLY appeals to the ONLY possibility currently available and the Evolutionist complains of our putting God in that gap. However, they have NO explanation at all to put in the Technetium gap in the Table of Elements. They again simply say, "It happened. Moving on….." and never offer an answer.

Theses 45: Polonium radio-halos show that the Earth's base granites MUST have been instantaneously formed.

Speaking of heavier elements; Polonium is one. If one Google's the geologist Dr. Robert Gentry's Polonium Radio-halos research; we find an interesting story.

Most times when one looks into some aspect of the Creation vs. Evolution debate - you will find arguments for both sides and one must pick which makes the most sense. In this case, all you find are articles saying this man has found something no Evolutionist has found a way to explain away.

What did he find that doesn't just rock Evolution's boat - but sinks it?

Briefly, Polonium decays away by emitting radiation which stains the granite rocks it is found in. Various isotopes of Polonium have the same number of protons (making them all Polonium) but different numbers of neutrons (making them isotopes of Polonium).

Each isotope has a different half-life and emits a specific strength of radiation while it is that isotope. One can look at the circular stain(s) and tell which isotope was there. One form of Polonium has a very short half-life; mere fractions of a second! Yet it leaves a tell-tale stain in the base granite.

The base granite rocks that Evolutionists insist are billions of years old were supposedly formed from molten rock that cooled over millions of years.

If this were actually True, the very short half-life Polonium halos would NEVER be found in these rocks! They COULD NOT be since the rocks were supposedly molten. Even if they could be there, the halos would not be perfectly circular - but would be distorted by the still cooling rock as it shifted around. The fact that we DO find them - and they are perfectly circular and undamaged indicates that these rocks were NEVER molten!

If the rocks did not form slowly from liquid rock cooling, what other choice is there?

Hmmm… I know of a Book that starts out, "In the beginning, God created the heavens and the Earth…." He spoke and it was…. and here are the Polonium halos to PROVE it had to have happened in an instant. There is NO other account in the world EXCEPT the Bible's Creation account that matches all the actual evidence including Polonium halos.

Evolutionists at first embraced Dr. Gentry's discovery - because he was part of their 'club' - although silent up to that point. They even published his finding in their publications; because he did not draw the obvious conclusions in that work - that God must have Created the base granites instantaneously!

THEN they realized the IMPLICATIONS of his work! - and all the Evolutionists cheered Dr. Gentry! Ummm….. no. They did not.

In an instant he went from being the hero to the goat and he was summarily KICKED OUT of their club and his groundbreaking work which had been hailed as a breakthrough discovery dismissed and hushed up with NO better explanation put forward.

If you open fire on the Evolution Religion's boat, get ready for stormy weather. You will be thrown overboard without mercy. ALL must be sacrificed to protect 'the Theory'.

Theses 46: Too much Helium. I love balloons - especially those that are filled with helium and float in the air. Helium is a great element - if you breathe it in and then talk, it can make you sound like Donald Duck.

Helium is great for another reason - there is not enough of it in Earth's atmosphere for the Earth to be more than a few thousand years old. Even given the most generous factors for helium production and the least possible removal - the Earth's maximum age is only about two million years old; well short of the many hundreds of millions of years the Evolutionist needs to make amoebas into men.

Evolutionists KNOW this is a FACT proven by the RATE Project's very careful scientific methodology to the extent it can be measured by man. The tests were conducted by qualified scientists.

Yet the tyrannical leadership of the Evolution camp simply rejects any and all data that shines an unfavorable light on 'the Theory'. They will not even attempt to prove the data wrong or the methodology invalid. It is enough for them that the experiments were done by Creationists.

Evolutionists respond by simply doing what they do whenever they encounter something that contradicts their Religion - they simply dismiss it without conducting any experiments or testing of their own to show the Helium present is the right amount if Earth were really billions of years old.

They sometimes employ one other option - finding a 'scientist' who does not even do any experimentation to concoct some explanation - regardless of how preposterous - and then they embrace it completely.

The evidences are clear and compelling: Earth is young. It was Created. There IS a Creator God.

Theses Forty Seven thru Fifty Seven: Evidences from Observable Nature Prove the Earth is Young! Examples:

Theses 47: Niagara Falls erosion rate is too fast.

Evolutionists often tell me that Grand Canyon MUST be millions of years old because I did not see it erode away and I have no hard data indicating it could have happened quickly. I may not have hard data for Grand Canyon with regard to erosion rate, BUT there IS hard data for another massive geologic formation: Niagara Falls.

While looking around in an antique shop I found a bunch of old pictures that used to be mounted on a device that; when held up to the eyes - gave a 3-D view of the picture. In that stack I found a picture of Niagara Falls…. from 1902; so I bought it for a dollar.

The Falls were just as impressive then as they are today - but that picture shows the Falls STRAIGHT ACROSS! - NOT in the 'horseshoe' shape we see today.

The generally accepted erosion rate of the Falls backward is about 2 feet a year after 1905 and almost 4 feet per year before 1905. The reason for the difference in erosion rate is that much of the water has been diverted by a dam and hydro-electric project. Less water over the Falls equals slower erosion. Keep that point in mind.

IF Niagara Falls began right at Lake Ontario as it began to erode backward towards Lake Erie; its present location is roughly 7.1 miles from Lake Ontario.

IF the erosion rate before 1905 was constant all the time the Falls were moving to where we find them today; it would have taken about 9500 years.

Problem for Evolutionists: If all of Earth's major geological wonders have been happening for millions of years: Why hasn't Niagara Falls ALREADY eroded backwards and reached Lake Erie?

What if the Falls DID NOT start out at the edge of Lake Ontario? The Falls would then be much YOUNGER than 9500 years. Likewise, what if the erosion rate was HIGHER during some periods while the Falls were forming? Again, the Falls would be much YOUNGER than 9500 years. It also could have been a combination of the two; the Falls did not start right at Lake Ontario AND the rate of erosion was higher in the past.

According to literal Biblical chronology; the Flood of Noah occurred about 4200 years ago. The ENTIRE Earth was flooded according to this account. Therefore, the volume of water that would have cut Niagara Falls in its earliest stages would logically have been MUCH higher than what we see today.

Do we have any 'hard data' that shows what the effects of changing the amount of water flow over the Falls has on the erosion rate? Yes, we do. Prior to a large portion of the water flow being diverted in 1905, the erosion rate was about 4 feet per year. After the diversion, the lesser amount of water flow eroded the Falls at about 2 feet per year.

It is LOGICAL to conclude that if the water flow was much higher as Noah's Flood initially drained into the newly formed oceans - the Falls would have eroded much more per year.

Therefore, we CAN get to the Bible's claimed age of the Earth by logic and reason. There is NO WAY to get anywhere close to the length of time Evolutionists need to make a microbe into a monarch butterfly.

Let us apply this reasoning to Grand Canyon - which will be covered again in more depth in a later Theses. That massive canyon we see today CANNOT be judged regarding age by the small river running through it now.

How fast can water work?

California in February 2017 is finding out. The State had been in a severe drought for 8 years and most reservoirs were nearly empty. In just a few days - a series of rainstorms provided enough water to fill to overflowing one of the largest reservoirs - the one behind the Oroville Dam.

This dam's spillway is.... or was: 1.3 miles long....

To control lake water level; excess water was let out through the normal spill way - some 45 million gallons per minute at peak release;

roughly equal to the volume passing over Niagara Falls. Peak water release lasted only about 4 days.

Apparently there was some defect in the spillway about 2000 feet down the concrete path - and the water began to cut through the cement.

Within hours a huge hole had been dug - severe enough to cause the humans to shut off the flow. The water then found its way to the emergency spillway which had never been used in the history of the dam. Within hours, it also had been eroded so badly that the cities below the dam were mandatorily evacuated!

Humans shut off that flow and the water again went over the main spillway and resumed cutting the previous hole. It has now been a few weeks and again the main spillway has had flow stopped - but there now is nothing left of the manmade spillway below where the first cutting began.

When the water stopped after just four days of peak flow, a gaping hole some 250 feet long, 75 feet wide and 45 feet deep had appeared. This equates to almost 94,000 cubic yards of rock! Let's do some math to see how long it would take the Oroville dam's spillway water to remove the amount of material removed to form Grand Canyon.

Grand Canyon is a big hole! Its volume is about 5,450,000,000,000 or 5.45 trillion cubic yards. The Oroville dam overflow removed an average of about 25,000 cubic yards each day it was at peak flow. If this flow had persisted, if would have taken about 597,000 years to remove the volume of Grand Canyon. Notice the water would NOT take millions of years to do this as Evolutionists love to CLAIM.

The Colorado River's peak flow has been measured as 3.8 times the peak flow experienced at Oroville. Therefore, we must divide the 597,000 years Oroville level flow would have taken by 3.8; which equals just under 150,000 years for the Colorado at max flow today to remove ALL the material to cut Grand Canyon!

This is IF the relationship to peak flow is linear! There is every reason to believe that the higher the peak flow - the rate of material removal rises exponentially!

Once again we see that the Evolutionist's presumed 'millions of years' are just a fabrication to try to discredit the Bible's timeline. If we assume Noah's Flood and the breaching of a natural dam resulting in the rapid draining of a lake three times the size of Lake Michigan is what really cut Grand Canyon - we can EASILY get that 150,000 years down to a few months worth of water flow happening about 4400 years ago.

The Bible timeline COULD still be True. Evolution's timeline has NO chance of being true.

Just like with so many other aspects of this debate, we can use logical assumptions to show the Bible's account is plausible. At the same time, our calculations - being as generous as possible to the Evolutionist side - grossly fail to yield anywhere near what Evolutionists have claimed for decades and state as 'fact'. To get to the claimed ages of Evolutionists, we must leave the world of reality completely.

If the Oroville dam had failed, the reservoir would have drained in a few days with multitudes more water flow than what was released through the spillway. It would have cut a new canyon and left a small stream behind.

Men who did not know of this catastrophic event, but believing in Evolution's millions of years might come along years later and look at the valley with small stream. They would then tell us that this process must have taken many millions of years. They would have found the remains of the man-made dam - i.e. OOPARTs - and simply dismissed this evidence because it did not fit their presupposition.

ONLY when shown the minute-by-minute video of what actually happened MIGHT they reluctantly admit they were wrong. I have met some die hard Evolutionists that I doubt even seeing the video would shake their faith in the Evolution Religion.

Look folks: GOD HAS THE VIDEOS of Creation Week!

Remember! If man can do something like build Oroville dam, God can do infinitely more. If we humans have figured out how to make recordings of events - God certainly has the viewable PROOF that this world was in fact Created. He does not provide that to us yet because He demands that we BELIEVE His-Story; the Bible - as written and originally intended.

Evolution is a RELIGION that can only be believed - and must be believed IN SPITE of the real, logical, reasonable evidence and analysis.

Theses 48: Some may say that the argument about the Falls above is invalid because it analyzes but a single location on this vast Earth.

In a moment we will travel to the other side of the world from the Falls.

First: I would DISAGREE with that assessment because if an all-encompassing Theory of Evolution is actually True; there should be NOT ONE SINGLE item across this earth that could not be adequately explained in such a way that it would 'fit' nicely into that explanation.

We have now looked at 47 Theses; EVERY SINGLE ONE fits Creation as an overall explanation - and at the same time completely contradicts Evolution.

Bible believing Christians need to demand that Evolutionists do what they demand of us. They must: Explain every single item in the entire Universe adequately in context with their 'Theory'!

All they can point to are THREE things that can be force-fitted into their paradigm: Circular reasoned Geologic Column with Index Fossils, Radiometric Dating that MUST rely on questionable assumptions and Majority Opinion of the 'scientific community'. Their entire 'proof' is but a three-toothed paper tiger. It growls ferociously, but that is all.

Stand your ground, Christian! Their emperor is naked!

Off to the other side of the globe for a look at the Great Barrier Reef off of the coast of Australia. It is the largest reef in the world and it grows very slowly - so slowly that many assumed it was many tens if not hundreds of thousands of years old.

Atheists have used this 'fact' to say to Young Earth Creationists (YECs) that the Bible's timeline cannot be true. They claim that if ONE item shows the Bible's timeline cannot be right that the same fact indicates we must accept the 'billions of years' explanation in its entirety. Turn about is fair play. In this book I am detailing just 95 of the many thousands of points showing Evolution is simply wrong.

Yet when I point out more than one problem with their claim for which they have no answer - somehow that makes me a religious zealot that can be dismissed without answering my challenges! They are not willing to play by the rules they set up for YECs!

I AM a religious zealot for Christianity and Young Earth Creationism! I admit it! Guilty as charged!

However, the MESSENGER's beliefs DO NOT allow dismissal of VALID evidence in his delivery bag!

By all means, shoot the messenger and throw his body in a ditch - the message he brought STILL has to be proved incorrect! Before you pull the trigger to silence a YEC like me: Remember what Jesus said to the Religious Leaders of His day regarding what the King will do to the wicked ones who not just reject His-Story, but also mistreat and kill His messengers....

As per usual with Evolutionists, no one had truly measured how fast the reef grew. Wikipedia says the reef is but 6000 to 8000 years old.

Many other sites say it is 25 million years old or more. ALL of the sites I looked at were at least neutral to Creationism, many were posted by an avowed enemy of Creationism.

Thousands or millions - who is right?

There is a main factor to consider: Has the growth rate been constant for the entire existence of the Reef - or has it varied? An article published in Nature on August 16, 1984 by Peter Isdale indicates a correlation between fresh water runoff and the growth rate of the Reef in that area.

Basically he found that higher runoff of fresh water resulted in higher growth rates to a depth of 10 meters. Higher growth rate would be INTERPRETED as greater age by one believing constant growth rate. Also consider that a large coral die-off seems to be happening over the last 40 years. This die-off would slow the growth of the Reef. A person arguing with my claim that the Reef is but 4000 years old may say that there must have been long periods of slow growth.

The main point is: No human KNOWS the actual age of the Reef.

Was there an Earth inundating Flood about 4400 years ago which tore up the entire surface of the Earth and would then mark the beginning point or birth of the Reef? Are there accepted studies and oft-referred to organizations that say the Reef is but a few thousands of years old? Yes. Even Wikipedia says the Reef is only 6000 - 8000 years old!

Conclusion: Since there are so many 'expert views' regarding the actual age of the Reef; we must conclude that it is actually of unknown and unknowable age to this generation of men UNLESS an EyeWitness saw its beginning and recorded the event that started it.

There is ONE Testimony of an event that COULD have started the Reef contained in a Book. That Book is the Bible's Genesis and that event is Noah's Flood - a world wide Flood that would have been violent enough to tear up every existing feature of Earth and redeposited fragments of still living coral in the place we now find the Reef.

There are those that HAVE ACTUALLY made studies of its growth rate have concluded from actual data that an age of just thousands of years is perfectly possible.

I was able to find ONLY CLAIMS of great age for the Reef, but not one actual STUDY showing any evidence as to why they concluded it was 'millions of years old'. The reason they claimed great age was simply its size - in their 'expert' mind; that alone required it to be millions of years old.

People: DO NOT get intimidated by those who simply regurgitate what they were INDOCTRINATED with by the Evolution Religion's high priests called professors. A claim based on nothing more than a so-called expert's opinion is NOT a sound foundation! Especially when that single data point MUST be made to align with the hundreds of other data points that argue for a young Reef.

Simple logic tells us that if hundreds of points speak to youth of the Earth and one (or a few) MIGHT speak to great age - it is likely that those few indicating great age are being misinterpreted!

Am I now appealing to majority opinion to convince you? In a way, yes. The majority of points having ACTUAL observable and repeatable DATA associated with them point to a Young Earth. When I appeal to majority opinion - it is not simply a consensus vote tally like the other side uses.

YOU still have to THINK and make up your own mind.

Theses 49: Niagara Falls in the United States, the Great Barrier Reef in Australia - now we travel to Africa and the World's largest desert: The Sahara Desert.

Deserts grow if the prevailing winds blowing across them take their searing heat and dryness over land that supports vegetation. The Sahara Desert is growing today and - if nothing changes - will one day end at the Atlantic Ocean. One study on it's growth rate indicated this desert began about 4000 years ago….

A missionary who worked in that area and was driving across it gave me a SEDIMENTARY rock slab with SEA SHELLS in it - found in that desert. Therefore, at one time, the Sahara was underwater. Hmmm… seashells on top of mountains and in the middle of the world's largest desert - naw, can't assign significance to that…. OF COURSE we can!! The Earth WAS completely FLOODED at some point in its past!

The question now becomes, when did this land dry out and when did the desert begin growing?

If you Google 'How old is the Sahara Desert?' - most of the articles will claim it is between 3 million to 10 million years old based on INTERPRETATION of SOME of the evidence.

However, it is a known FACT that people lived in regions now made uninhabitable by the harsh desert climate. These people - known as the Garamantes - were not nomadic, but had formed settlements. These settlements seem to include gardens.

If the desert was actually at least 3 million years old - as old as the oldest claimed human evolutionary ancestor - while man was still little better than a chimpanzee that perhaps could use a few crude stone tools - this people group would NEVER have been able to establish these settlements.

The British Broadcasting Company's science editor Dr. David Whitehouse reported on German scientists who were using a new computer modeling program to figure out when the Sahara was born. Their model showed that there had been a 'brutal climate change' - NOT due to human land use - where rainfall had essentially stopped. WHEN did this occur? Hmmm…. about 4000 years ago…. which would have been about 400 years AFTER the Flood of Noah….

The article goes on to say that the peoples living there in settlements would likely have been driven to other water sources…. like the Nile…. Hmmm…. Genesis tells us that Abraham fled a drought and (about 400 years after the Flood) went down from Canaan to Egypt….. near the

Nile….. The Flood of Noah is best estimated to have occurred about 4400 years ago… Hmmm….. must be just a coincidence…..

When ALL the evidence is carefully looked at; a very CREDIBLE idea takes shape that just happens to MATCH in great detail the account found in the best selling Book of all time; The Bible.

We still must BELIEVE the Bible is True, but there is every reason to do so.

Theses 50: Speaking of deserts, about 15 years ago I was visiting one in the United States southwest: The Sonoran desert to which the Saguaro cactus is a native species. It will live elsewhere if transplanted, but is native to this one place in the entire world. If Evolution is true - it MUST have evolved there.

Evolutionists tell us that this process of Evolution takes millions of years for one type of living thing to change into another.

If you ascribe to the 'Super Amoebae' claim - the great Saguaro cacti came from said super amoebae. Hmmm…. an amorphous single-celled blob that lives in water eventually having an off-shoot that developed into a plant with SPIKEY spines able to thrive in a very dry and hot desert…. yeah, right.

If Evolution is really this powerful - it is no wonder humans bow down in worship to 'The Theory'! However, thinking persons will exclaim, "Houston, we have a problem."

You see, my family took the tour given by the National Park Service. The young guide of about 22 years old told us that the desert is actually relatively young - on the order of only about 10,000 years. Before it was a barren desert, she told us that it had been a lush grassland. Why did she say this?

Numerous mammoths and other critters are found fossilized there. She had previously stated that the elegant and towering Saguaros are native to this desert only - out of all the world's deserts.

Not wanting to be mean, but honestly wanting to hear the answer she had been indoctrinated with; I asked, "Then how did the Saguaro come to be? It needs this harsh desert environment to thrive, yet only 10,000 years ago that environment did not exist."

She answered with what she had been told to answer, "The mighty Saguaro EVOLVED from a small shrub that lost it's leaves when the desert conditions arrived. It bulked out its trunk to conserve water."

I did not want to embarrass her, so I let it go at that - but obviously this is merely another Evolutionary fairy tale.

Let's see if there might be an account that fits better and does not make my brain hurt:

God created Saguaros as a hardy species. It was one of the unique plants in the Garden of Eden. Noah's Flood destroyed everything from the original Garden and many plants managed to survive by going dormant until the water abated. Then these plants grew - if they could - in their new environment. This would allow us to find unique plants in unique environments all over the world as well as see only fossils of extinct plants that did not survive the Flood.

What do we actually see? Unique plants in unique environments all over the world and fossils of plants that did not survive the Flood.

Folks, Creation works every time! No mental gymnastics or wild 'just-so' stories needed!

Theses 51 and 52: I have said whatever Theory is actually True MUST be True and logical all over the world. Let us go to the Arctic Circle next to see if Creation works there as well.

Mammoths were roughly the size of an elephant. Modern elephants each eat at least 200 pounds of food a day - it is logical that a mammoth would have eaten a similar amount daily.

National Geographic Magazine's April 2013 issue indicated there are 'millions of mammoth tusks' above the Arctic Circle. Millions could be 2 million animals of this size - so let's just say this environment would have had to produce at least 400 million pounds of edible vegetation on a sustainable scale so that each and every DAY this massive population of giant plant eaters would be fed. This volume of plants is just to support the minimum population we know of; other estimates say that as many as 5 million mammoths lived up there!

Nothing much grows there today of the sort or on the scale that would come anywhere close to being able to sustain even the minimum estimated herd of these animals. Combine this with the fact that our oil drilling rigs north of the Arctic Circle have brought up TROPICAL PLANTS from great depths….. and a thinking person would conclude that the world was a very different place when these millions of mammoths were alive.

Many of the mammoths have been found frozen in the standing position! They did not die of a heart attack and fall over; neither did they drown - but were flash frozen ALIVE with their last meals often found in their teeth!

All of the evidence - went rightly considered - indicates some kind of worldwide catastrophe happened that was capable of ripping up millions of square miles of plants and turning them into coal, depositing almost pure layers of rock hundreds of feet thick, and killing by suffocating and then FREEZING millions of critters eating summer-time plants as it transformed the area above the Arctic Circle into a permafrost wasteland.

Mammoths are also nearly always called 'Wooly' and are almost exclusively shown in a deep winter scene with huge snowdrifts and NOTHING for them to eat anywhere in sight. This scene is just an artist's conception.

The analysis of mammoth and other animal carcasses indicate that the mammoth could withstand cold weather and that their bodies decayed somewhat before being completely frozen - but we MUST NEVER allow a label or name to convince us of any critter's manner of life or death. Just because these giants were 'woolly' DOES NOT mean they lived exclusively in the freezing Arctic of today.

Mammoths are not found JUST in Arctic areas.

Remember that my Sonoran Desert tour guide indicated mammoths had been found in what is now southern Arizona. They have been found in Florida, Texas, Michigan, Indiana and many other temperate or even tropic places like Mexico and Patagonia.

Folks - it is time to think for yourselves: Evolution requires SLOW changes over millions and millions of years. Yet our Earth is screaming out to us that it has suffered a near fatal catastrophe! Evolutionists say it was a meteor strike about 65 million years ago - because even they finally could not ignore all the evidence we YECs put before them.

Yet, they cannot bring themselves to even consider God or entertain the idea that God might be the God of the Bible and that Genesis is His-Story. They had to come up with something to kill off the dinosaurs and reconfigure the world - so they imagined an asteroid strike did it.

However, the event that killed off the mammoths was only thousands of years ago - for these animals are often not fossilized or found in layers dated to millions of years old by Evolutionists. If you are an Evolutionist, you say it was some unknown event at the end of the last ice age that took out the mammoths. If you are a Creationist, you say it was Noah's Flood that did ALL of these things.

Everything - yes, literally EVERYTHING - fits very nicely with Creation. Evolution is internally and fatally flawed - what you believe about one thing contradicts at least one other thing and often many other things. Go with what works every single time: Creation Theory as told to us by God the Creator in His-Story.

Theses 53: Some things live to be very old. Trees live a long time, too. In California, the oldest known tree (by coring and ring determination) is almost 5000 years old. Using rings to date trees can give an indiction of the tree's age, however it is known that a tree can add more than one 'annual ring' each year and therefore appear older than it is. Another tree is claimed by its champions to be some 9500 years old, but it was aged using the already discredited method of radiometric dating.

Question: Why is the OLDEST known living thing on land ONLY about 4400 years old. To say the California bristle-cone pine dated to about 4700 years old is actually only 4400 years old; we must assume that a few years occurred where multiple rings were added in a single year.

This is a reasonable assumption considering that we have observed this phenomenon of multiple tree rings in a single year today.

If Noah's Flood tore up everything and a few bristle-cone seeds survived the Flood, finding a place to grow - then we would EXPECT the oldest living thing to be in this age range.

Main point: We DO NOT find living things on land older than this range EXCEPT if we allow discredited radiometric dating to be used.

Theses 54: The comic strip B.C. was famous for saying, 'Clams have feet!' They would have to have feet since clams and many other examples of fossilized marine life were found on top of Mount Everest. Marine fossils have been found at the tops of every major mountain range around the world.

The obvious conclusion that a thinking person would come to is that the ENTIRE EARTH must have been covered by water at some point in the past.

There is an account in a Book that says this did in fact happen. According to the Hebrew word used in the Genesis 6-9 account, the mountains of Noah's day were likely on the order of a few thousand feet at most. The mighty mountains of our day are the result of the 'breaking up of the great deep' and the other geologic catastrophes that followed the initiation of the Flood. We are still experiencing some of these affects today - but on a much smaller scale.

It is therefore perfectly legitimate for Creation believers to not be surprised at all that marine fossils would be found in places Evolutionists have a hard time explaining.

Theses 55 and 56: The products of erosion MUST come from somewhere and go somewhere. The dirt comes from the mountains and ends up in the river deltas.

Evolutionists do not like to talk about these two theses much - probably because their own data shows that the mountains are eroding far too quickly for their standard 'tectonic plate theory' to be valid. At the current known SLOW rates of erosion - the continents should have already eroded flat about 10 million years ago.

That being said, in the next breath these same Evolutionists will claim that fossils of critters found IN THE ROCKS of these 'should have already been eroded flat' mountains are hundreds of millions of years old!

I said previously that Evolution Theory is self-contradictory and oftentimes self-defeating; as is the case here. They simply cannot have it both ways. Either the measurable and repeatable science of erosion rates is wrong - which means their geologists are wrong about things they CAN SEE and CAN MEASURE - which means if they cannot get this right, why believe them about anything else - OR - the fossils we do find on these mountains are actually FAR YOUNGER than they tell us. Period.

Now that the dirt is washed off the mountain and is in a river; eventually it winds up in the ocean at the mouth of the river; i.e. in the river's delta. If the process of erosion had been going on for as long as Evolutionists claim - there should be a LOT more dirt at the deltas of the world's major rivers. There is simply not enough dirt in these locations for the erosion process to have been going on more than a few thousands of years.

Again, this is using TODAY's slow rates of erosion! Much of that delta dirt was likely deposited within the first few years after Noah's Flood. At the base of the Oroville Dam there is now a natural dam hundreds of feet long that has sealed off the discharge of the hydroelectric plant! - and this happened in less than ONE WEEK of peak flow!

Even if an Evolutionist tries to claim that based on today's slow rates this delta is 50,000 years old (implying the Bible is not True) - he forgets on purpose to take into account the Flood. God reminds them of the Truth of His Word by allowing situations like the Oroville Dam or Mt. St. Helens to occur.

Do you see why after more than 30 years of discussing every possible aspect of this debate with Evolutionists and Atheists - that there is not even a single, solitary shred of evidence that is not better explained by Creation concepts?

I believe Creation BECAUSE of the actual evidence. They believe Evolution IN SPITE of the actual evidence.

Theses 57: Time is NOT the friend of nothing-to-something-to-molecules-to-man Evolution. Invoking vast amounts of time is simply a PLOY to make something that IS IMPOSSIBLE seem to be possible.

Though we are continuing to beat a dead horse; these previous few Theses each have dealt a fatal blow to the already dead Religion of Evolution. If any of them show that there never has been 'billions and billions of years'; then even a 5 year old can see that worms do not evolve into wombats or women.

The reason time is NOT the friend of the Evolution Religion is because during every second of every minute of every hour of every day of every month of every year of every century of every millennia - the Law of Entropy is working far more effectively to decay everything than the Theory of Evolution's unknown organization process could ever overcome for even a brief period of time in a very localized place.

If you want to get an Evolutionist really angry at you - challenge him over the issue of how long the Earth has existed. They will literally go into orbit insisting that the unobserved 'billions of years' are a fact and call you every derogatory name in the book (and invent some - like Creatard; a Creationist Retard or IDiot - an Intelligent Design idiot).

The decaying effects over time is actually an evidence that there IS a Creator God that is STILL INVOLVED with His Creation. You see, without Intelligent intervention by a Sovereign Creator, this world's complex systems that make life possible at all would never have come together all in one place all at one time or lasted even these few thousands of years.

If you think about it, **PRAYER** is **ASKING** the Creator to become involved to stop or reverse one or more of these decaying effects which are the natural consequence of the Law of Entropy.

This is WHY God told the Jewish people to look for the Messiah - and gave them a way to identify Him when He came. He would be 'the man of miracles'!

Nicodemus admitted as much when he visited Jesus saying, 'We KNOW that you are sent from God, for no man could do the things that you do.'

Yet, because Jesus threatened the Jewish leader's power and position - they labeled Him a fraud and killed Him. Modern day RELIGIOUS leaders of the Evolution Religion would like nothing better than to silence Creationists permanently. This is why absolutely NO challenge to Evolution's monopoly in public schools is tolerated.

However, the unjust murder of the Messiah some 2000 years ago was allowed to happen so that the ULTIMATE MIRACLE could be done right before their eyes!

Jesus ROSE from death!

He DID what the Evolutionists can only fantasize about - a pile of dead chemicals and what little water had not evaporated out of the corpse - came to life!

The Bible contains the ONLY actually True examples of abiogenesis. There are several accounts where something known to be dead came back to life - after God created men from dust in an instant the first time!

In all those accounts - Elijah raising the widow's son, Elisha raising the Shunnamite woman's son; Jesus raising Lazarus, Jarius' daughter and a different widow's son; the dead saints rising after Jesus' resurrection and appearing to many in Jerusalem; Peter raising Dorcas; Paul raising the boy who fell out of a window: ALL these examples of the physically dead coming back to physical life were done by others for others.

There is ONLY ONE example of a Person raising THEMSELVES back to life from death.

That example is Jesus.

He PROVED Himself to be the Son of God because He claimed He had the power to lay his life down (physically die) AND to take it back up again! Then He SHOWED us by dying the most brutal death imaginable - staying dead for three days - then RISING to Life Eternal!!

He was seen by more than 500 people in the days after His execution. Professional prosecutors know that one NEVER makes the claim that they could interview some 500 people and get the same basic testimony UNLESS it was actually True. Yet the writer of the Book of Acts makes this claim - and this author was a doctor, an 'educated man' and one very familiar with the science of his day.

This IS the ultimate Testimony that there IS a God AND that Jesus is One of the Triune Godhead. There is NO other way He could be alive after being killed.

This is also why Atheists fight so hard against the Gospel - if it be True, then THEY are condemned for not believing it.

Theses Fifty Eight:
The Effects of Salt on Existing DNA Strands

Premise: Since DNA can be 'unraveled' by mixing its tightly packed strands in a solution of water, chemicals and salts; then it is logical to conclude that DNA could never have been compacted into its currently found form if it 'arose' in a 'warm, salty pond'.

Theses 58: Most people brainwashed by the MYTH of Evolution fully understand that life from non-living chemical soup is impossible. Therefore, they ALWAYS insist that I let them start with their mythical first super-cell, even though when I press them on the issue - they will often admit they have no clue how such a cell could have come about.

True-to-form, they simply BELIEVE that starting with a living cell is a valid position - and they will completely ignore my attempts to get them to think about how impossible it is for such a cell to come to be.

One thing I point out to them is that DNA is the most complex and tightly compacted string of information known to man! DNA is far smaller than our most advanced super computers. It runs on its own, often repairs itself and has a complete SECOND copy to work from if repairs cannot be made to the first copy! Each of the 25 TRILLION cells in your body (except red blood cells that have no nucleus) have this dual copy of YOUR unique DNA.

This DNA can be replicated by those cells that divide. The cell makes an entire NEW dual set of DNA for the new cell! It does this automatically and correctly for most all forms of life on Earth.

Yet the Atheist in his absolute pride and arrogance simply dismisses this work of God and credits it all to a chance reaction of dead chemicals in a '…warm, SALTY pond…' (and do not forget great ages of time in this recipe) '…billions of years ago…'

We've already punched numerous holes in the millions of years claim, now let's take a quick look at a simple science fair experiment: Unraveling tomato DNA.

There are various recipes to do this, but the point here is that you use a 'buffering solution' at one point in the process.

One website's buffering solution is a liter of water, about 4 grams of sodium chloride (salt) and sodium citrate (another salt). As described in another website's methodology - this buffering solution is used to 'neutralize' the electrical charges that hold the DNA tightly bound in a microscopic form. This is done so that a 'white, snotty looking' product becomes visible. This 'white snot' is the unravelled tomato DNA.

The point is this: SALTs neutralize the very electrical charges that allow DNA to be tightly bound in a microscopic cell. Hmmm…. 'the first cell evolved in a warm, SALTY pond….' - no, not possible.

Once again, Evolutionists will zero in on any ONE THING that sounds like it might work - declare it as undisputed fact - deride anyone and everyone that points out the fatal flaws in their fantasy - and simply move on to brainwash the next generation of school kids - at TAX PAYER expense I might add…. again.

That brainwashing occurs was pointed out earlier when I recounted what was found in the seventh grade biology 'science' book. On one page it gave real science regarding Louis Pasteur proving the Law of Biogenesis and proving wrong the 'Law of Spontaneous Generation'. On the VERY NEXT PAGE the topic of how life 'arose' on the 'early Earth' was stated as known fact when the very process they described is Spontaneous Generation that NEVER would have resulted in the microscopic DNA necessary for cellular life!

Theses Fifty Nine to Sixty Six:
Not a 'Missing Link' - the Whole Chain is Missing!

Premise:

If man really 'arose' from a chimp-like critter; the fossil evidence should be:

1) very clear and compelling - not fragmentary and able to be interpreted several different ways

2) the details surrounding the finds should be without suspicion, and

3) the people interpreting the finds should not be able to be contradicted or contradict themselves

Therefore if we find:

1) questionable evidence (or are not allowed to examine it at all, ever) and / or

2) find suspicious details about the find itself and / or

3) we find actual, provable FRAUD with intent to DECEIVE the public

- THEN we are justified in DOUBTING those finds and 'X-ing out' those critters in the supposed chain from monkey to man.

Theses 59: Evolutionist's famous depiction of a series of creatures slowly going from crawling on all fours to walking hunched over to carrying a club and then a spear; walking upright to a 'modern man' is well known to anyone who has attended public school or for that matter has eyes. The purveyors of Evolution are master marketers and have posted this paradigm altering picture in numerous venues.

The problem is when we look at the 'actual fossils' and the circumstances surrounding their discovery we find:

1) missing or reinterpreted evidence
2) outright fraud committed
3) the Evolution community in disagreement over the find, and
4) evidence not allowed to be examined by a competent person that holds a view other than Evolution

Since this book was not meant to be an exhaustive treatment on any one subject; we will only look at some of the chain; enough to see that this mental picture which indicates a clean line of ascent from monkey to man is nothing of the sort. The ONLY place this 'complete' chain exists is in the fertile imaginations of Evolution Religion believers.

Theses 60: Lucy: Digging in some desert or the depths of some awful location for bone fragments must really suck. It might hold some excitement for about a month, but then the poor sap who spent tens of thousands of dollars on his otherwise worthless Anthropology degree realizes that UNLESS he finds SOMETHING to take on the lecture circuit - this is his life: Heat, insects, probably dangerous animals or people, dirt, tents and cots, outhouses… no fame, no fortune, no adoring followers.

There is another pressure as well: Grant money. This free money ONLY goes to those who PRODUCE the desired results for the DONOR of the money (Soros?). It does not matter how contradictory the OTHER evidence found with the main find is - if you do find something, you MUST interpret it in line with the Evolutionary belief of the donor; even if that means ignoring contradictory evidence. This is why so many really marginal fossil finds have been hyped so much; like Lucy.

Let's compare Dr. Donald Johanson's situation today versus when he made 'the find' of a lifetime: Lucy. He was out in a desolate place living in tents - no running water, no comfortable bed, nothing but dry landscape and rocks. There was no fame and certainly no fortune at this point. The Leakey family was getting all the glory… and on the speaking circuit; getting paid to regale their audiences with details of their finds.

One can imagine that the good doctor realized that the ONLY ticket out of this awful place is to find something, anything AND de-thrown the reigning 'oldest' pre-human.

Wallah! - he finds some fossil bones and begins to trumpet the news.

Fast forward; he lands a gig for some $250,000 a year in an air-conditioned office at Arizona State University - and regularly gets paid to recount his singular exploit and write books.

WHAT exactly did he find?

Let us take a look at Lucy. The skeleton is only about 40% complete. Analysis by other Evolutionists cast doubt on just exactly what it was. One can find many - Creationists and Evolutionists alike - that say it was just an extinct form of tree-dwelling ape.

ONLY those who have convinced themselves that one type of critter can change into another, different type of critter count Lucy as a 'human ancestor'.

There are many claims that Dr. Johanson said in speeches early on that parts of the skeleton were found in two places. This claim is disputed, but one must take into consideration the FRAUDS perpetrated by numerous other Evolutionists and detailed in the Theses following this one.

One thing is for sure; there is too much controversy and disparate interpretations of 'Lucy' - on the Evolutionist's side - to rely on this fossil and its interpretation.

After all, we ONLY have Dr. Johanson's word regarding how and where he found Lucy. Even if he recorded every detail of the find accurately, the account STILL is from a man who believes things that are provably false - that creatures change from one kind into another and that the Earth is billions of years old and that radiometric dating is reliable.

Atheists and Evolutionists constantly belittle the Bible because it was 'written by fallible men'. Turn about is then fair play - for, check me if I am wrong, but Dr. Johanson IS also a fallible man who has written about Lucy.

Hmmm… believe a fallible man or believe God Who wrote His Account using fallible men (and He alone is able to override their fallibility long enough to pen His Word).

It seems that no matter which alternative I choose - I am forced to believe the writings of fallible men.

Given this choice…. ummm…. I am going with God after a good look at ALL the evidence shows clearly that every piece lines up with the Creation account.

Theses 61: Fluffy the cat. Before we delve further into the other claimed ancestors of man, let us get one thing straight: The ONLY thing Dr. Johanson, the Leakeys or any other Evolutionist EVER finds are fossilized bones, footprints made in mud turned into stone, stone tools and OOPARTs (but they ignore the OOPARTs or dismiss them out of hand).

They NEVER find a written account from the fossil critter.

Therefore, they can ONLY make up a story to describe what they THINK may have been. They then rely on a smokescreen of 'PhD this and Professor that' to get folks to accept their interpretation. The old 'trust us, we are the experts' line.

To illustrate this point, may I tell you about Fluffy?

Fluffy was one of those nomadic neighborhood cats that no one claimed as theirs, but enough fed Fluffy so it never left. Tragically, Fluffy got hit by a car and was killed. Since she was no one's cat, no one picked up her body and it decayed on the roadside, after a few days becoming nice and dry and crispy in the South Carolina summer heat.

Finally, I had a brainstorm!

I went over to where what was left of Fluffy lay - some bones and teeth, dry skin with fur and dried up insides. I carefully interred Fluffy's remains in a sealable clear plastic box.

Why?

Years ago I used to give presentations on Creation vs. Evolution and was looking for a way to show people just how little a Dr. Johanson really could KNOW about the critter that once had the body belonging to the bones he found.

After holding up the box and letting them see what was in it; I would ask for a volunteer from the audience to answer some questions about Fluffy. I would usually pick the most squeamish teenage girl so that there would be plenty of 'Ewwwws!' - hey, it was good for a laugh. Be sure, you MUST entertain an audience if you want them to pay attention these days.

I would ask the volunteer to identify the animal and they would look it over and say it was definitely a cat.

I would congratulate them on their answer and then ask them what Fluffy ate? They invariably say, 'Cat food.' to which I would followup with, 'Do you KNOW this - or - do you simply believe this because of YOUR experience with cats eating cat food?' They would admit there was no way for them to know what Fluffy ate - heck, I did not even know, because we never fed Fluffy and I personally never saw her eat - and I saw her alive for months.

Notice, even I made an assumption in the opening account of Fluffy! Did you catch it? I assumed that the neighbors fed her. I never saw them do it. ALL OF US make assumptions - including those finding bones in deserts....

Then we would go deeper. I would ask, 'Did Fluffy have any kittens?' - the volunteer would say, 'I don't know.' I would then ask if Fluffy had any mutant off-spring?' They again would say the obvious, 'How should I know?' I would then have them sit back down to make my final point.

You see, folks, if I never actually SAW a creature alive and KNEW from first-hand eye-witness accounts the details of the animal that once had the fossil bones inside their body; there is absolutely NO WAY for me to KNOW for a fact ANYTHING about the creature. All I can do is take the best GUESS based on all the evidence available such as other animal fossils or fossil plants or the layer of rock the critter was found in.

Dr. Johanson can no more KNOW anything about how Lucy lived than I could about Fluffy - and I SAW Fluffy alive! I never saw if she had kittens or if she ate only cat food - for in South Carolina; outside cats also eat moles, mice, lizards and grasshoppers. I do not even know if 'she' was a 'she'! I just picked a gender!

Likewise, ALL Dr. Johanson or ANY other fossil finder could do is say 'I found this bone here and that bone there, these are the other fossils I found nearby in the same layer', etc.

From a few crushed skull fragments one MIGHT be able to take a shot at how big her brain was - but that again in NO WAY indicates how intelligent or not she was.

As you read about these next so-called 'links in the chain' from ape to modern man - remember what can be known and what ONLY be guessed at.

Theses 62: Nebraska Man - the pig that made a monkey out of all Evolutionists!

In 1922, a single fossil tooth was examined by an Evolution Religion believing scholar. Though worn, he quickly claimed it to be a possible primate relative of man. From this ONE TOOTH, an artist rendered a complete drawing of a primitive man AND HIS FAMILY!

Later excavation proved this tooth was actually from a type of PIG!

While the Evolution defending TalkOrigins website tries to claim this find was not significant - if you X-out both Lucy and Nebraska Man from the aforementioned drawing showing man evolving from a monkey - we now have at least three 'missing links'; the original 'missing link' plus the dubious Lucy and mistaken (or fraudulent) Nebraska Man.

I do not know about you, but I have never had much use for a chain with one missing link, let alone three missing links. Let's move on - SURELY there is much more and better examples of 'pre-humans' available and we will be able to mend this chain and conclusively show that man evolved!

Not so fast....

Theses 63: Java Man - Dr. Dubois found a skull cap, femur and three teeth - but wait a minute...

Two of the teeth have now been determined to have been from an ape and the femur has been determined to NOT be of the same age layer as the apparently far-older skullcap.

Hmmm.... all there is of Java Man then is a skullcap and ONE tooth - and even these are questionable regarding them belonging to the same individual due to the distance between the fossils.

Yet Evolutionists trumpeted the news of the 'missing link' between apes and man had most definitely been found.

Never mind that Evolutionists built another entire being from such scanty evidence and never mind that the guy who found it had announced that he was setting out on a mission to find 'the missing link'. A little bias on the part of the good doctor? Ya think?!

Again consider that even if Dr. Dubois was able to set aside his bias and did find everything exactly as he recorded - we STILL MUST accept or reject his INTERPRETATION of the fossil.

There is some evidence that even Dr. Dubois may have reconsidered his own interpretation, though he apparently never recanted his view that he had found the Holy Grail of Evolution.

Remember folks - this man believed one type of critter could change into a completely different type of critter. This has NEVER been demonstrated to be true, it can ONLY be believed (religiously).

If you want "Just the facts, ma'am" - an honest person must at least put a question mark over Java Man, but would also be perfectly justified in X-ing him out of the chain as well.

Theses 64: Piltdown Man - an outright FRAUD fools Evolutionists for decades. They wanted so badly for the missing link to be found that they accepted the claim with NO objective examination.

A relatively modern skull with an ape-like jaw. Found in the early 1900s - not exposed as the fraud it was until forty years later! This 'find' had been taught as THE missing link in colleges to millions by the time it was outed!

Again, a complete 'understanding' of this 'important part of the human history' was created out of whole cloth!

Should evolutionists be guilty by association? Generally, no one should be judged guilty just because they are associated with SOME bad actors. However, given we have YET to see a SINGLE unquestioned 'human ancestor' find coming out of the Evolution camp; it IS valid to henceforth regard WHATEVER any Evolutionist says about a bone or skull fragment with suspicion.

Lucy proclaimed at first to be a pre-human; now regarded as suspicious and even denied by many Evolutionists.

Nebraska Man hailed as a pre-human by many Evolutionists; now known to have NEVER been any type of primate or hominid.

Java Man hailed as the missing link, now considered an unusual and indeterminate find.

Piltdown Man a complete and utter fraud that fooled Evolutionists for decades.

Would someone please tell me WHY it is that I am looked down upon for choosing first to DISBELIEVE anything coming out of the Evolution camp unless it:

1) Has been thoroughly examined by Creation believing authorities and

2) Has no other explanation at least as reasonable as what the Evolutionists are claiming.

Given these two criteria - so far, we have neither a beginning to the chain, no middle and we have yet to discuss the end of the chain.

Atheists and Evolutionists claim there are errors in the Bible - yet none have been proven to be more than a possibly poor treatment of the translation to a different language from the original Hebrew, Aramaic and Greek.

You see, God's Word IS inerrant in it's ORIGINAL language as understood by it's ORIGINALLY intended audience. Therefore, when translating from an exclusively agrarian and animal based culture ruled by kings and such, half a world away - to - our modern English world of power split governments, electricity and cars - there will logically be some issues to be resolved by going back to the original Words and trying to understand it from the cultural setting in which it takes place.

When this is done, any supposed 'errors' disappear.

Even if they were 'true errors' - the ones claimed to exist by those I have argued with are really very minor and do not deal with major doctrinal topics.

Let me be very clear: I DO NOT agree with ANYONE that has shown me an 'error' in the Bible.

THEY maintain they have found at least one error.

Even if they were right (which they are not): NOT ONE supposed Bible error rises to the level of a total FRAUD that fooled the entire 'scientific world' for decades.

How many times do the Evolutionists get to be completely wrong or in utter disagreement with each other or caught LYING before I get to stop believing them?

The folks online tell me that because THEY think there are errors in the Bible that I must instantly distrust it completely. By their own standard applied to the 'chain from monkey to man'; I am completely justified in rejecting and even laughing at the entire notion of Evolution!

Let us move on to 'near modern man' finds - keeping in mind every major link in the chain thus far is either missing, a fraud or severely suspect to this point.

Stunningly: It doesn't get any better.

Theses 65: Neanderthal Man - gets his name from the place the bones were first found; the Neander valley in Germany. The German word for valley is thal - so his name literally means Neander-valley. It just sounded so primitive that the Evolutionists went with it.

Lets get right to it - the Evolution believing 'scientific community' has classified this creature as Homo Sapiens Neanderthalensis. We - fully modern humans - are classified by these self-proclaimed Wile E Coyote super-geniuses as Homo Sapiens Sapiens.

Let that sink in for a minute.

Some have said that if you took a Neanderthal, shaved his artist rendered beard and cut his artist rendered long hair, dressed him in a suit and let him walk down a street in New York City - he would attract no more attention than anyone else. In other words, he would LOOK just like us 'modern humans'.

Since his bones were found in a cave and there were stone tools found as well as evidence of fire - again, the wild imaginations of Evolutionists desperate to finish forging (pun intended) their chain from monkey to modern man created an entire society from the very fragmentary evidence.

One speaker I heard claimed that ALL of the major fossil bones from ALL of the major 'considered to be in the linage of man' finds - even the frauds - could fit into a single, standard sized coffin - with ROOM TO SPARE!

Think about that for a minute.

If you have only about 250 pieces to a billion piece puzzle - and some of those pieces may not even be from the right puzzle (but you do not know that) - how in the world could you KNOW with any certainty what the puzzle should look like?

You DO NOT have the box cover with the picture to look at! All you have to go by is IMAGINED pictures of what the completed puzzle MIGHT look like!

Given all this: There is no way for anyone to KNOW if the puzzle pieces are even from the right puzzle!

Sheer belief - sheer RELIGIOUS belief - would have to be substituted for factual knowledge - and has been when it comes to Evolution.

This would mean that the people who follow after Evolution and call it science are following after 'science, falsely so-called'.

Anyone familiar with the writings of Paul the Apostle will know the passage of prophetic Scripture to which I refer. He said that in the End Times a very powerful lie (Evolution) would arise that would be 'science, falsely so-called'.

We are here, folks.

Welcome to the End Times.

Theses 66: Cro-Magnon Man - this time named after a cave in France rather than a valley in Germany. These fully human skeletal remains resulted in the scientific classification of Homo Sapiens Sapiens.

WHAT?!

That's right folks - there is NO discernible difference between Cro-magnon Man and us.

Yet he appears in the 'chain of human linage' PRIOR TO 'modern man' to help fill out the imagined line.

Neanderthal could have passed for a modern man and he was said to be from about 50,000 years ago.

Cro-magnon is force dated to about 4000 years ago and is called, 'Early Man'. The tools, etc. found near him as well as cave wall paintings attributed to him are THE ONLY things that make him 'early'.

Excuse me?

I made and used homemade spears as a boy. I would find a likely stick and a sharp stone that I would attach to the stick and then go try to spear fish with it. It did not work very well. I was in Boy Scouts and part of our camping trips involved using 'primitive tools'. We would smash open nuts with stones, etc. There are even 'Paleo Camping Trips' you can take now days where you have to make your own tools.

Again we see that INTERPRETATION is the key to 'seeing what you want to be there' - as opposed to what is actually there.

They found BONES and what appear to be STONE TOOLS and some ARTWORK on some cave walls. THAT is what is actually there - THE REST of it is just their imagination.

By the way, there are cave men mentioned in the Bible in several places! The Gaderene Demonic was a man who lived in the tombs. In those days and that area, caves were often used as tombs. The traditional site where Jesus was born is a shallow cave. Elijah was living in a cave for a while after running from Jezebel. It was there that God visited him in a series of visions. The Apostle John would likely have lived in a cave when in exile on the Isle of Patmos.

Living in caves DOES NOT automatically make you primitive or ancient.

This section on the supposed Evolutionary chain from monkey to man deserves a summary statement:

There is NO BEGINNING to the chain - for it starts with a 'common ancestor' that has NOT even been found.

Every 'link' in the chain is at best comprised of a few bones and teeth - often in very bad condition.

Lucy is questioned by Evolutionists themselves - scratch her.

Nebraska Man was a pig - scratch him.

Java Man is now only a skull cap and one tooth - found by a man on a mission to find 'the missing link' so he was obviously biased - perhaps we can be generous and give Java Man a question mark (I wouldn't).

Piltdown Man was an outright FRAUD - remove him.

Neanderthal Man and Cro-magnon Man: BOTH are classified as Homo Sapiens - so any real differences between them and us are essentially in the minds of the evaluators.

People live in caves today. People lived in caves throughout the time period covered by the Bible.

I have visited many caves and could have camped in some of them.

I have made stone and wood tools and used them as a boy - yet I teach nuclear power technology today (i.e. I am not primitive or of low intelligence).

Folks - please understand that there is far more missing than a single 'missing link'. There is no beginning to the chain that has been found and the end links are fully human. Everything in between has been a fraud, misinterpreted or is in doubt even by the Evolutionists.

Case closed.

I do not care how many new 'finds' the Evolution High Priests announce to the world.

EVERY ONE will be just bones and possibly tools / paintings and THEIR BIASED INTERPRETATION of these items.

Period.

End of discussion.

Man was CREATED by a Loving God Who has provided a Way - a SINGLE WAY - for His created sons and daughters to have their sins forgiven and be reunited with Him.

That WAY is by the Blood of our Brother - the God-man Jesus Christ. Believe on Him and give God your life and you will be saved! Please, get saved now! Yes. Right NOW! Put the book down and pray to God to forgive you and save you.

He knows how you have been lied to and misled by those who follow the false Religion of Evolution.

We have another 29 Theses to go - and we will get to them - but remember: Tomorrow (or even later today) is NOT guaranteed to anyone.

Every breath you take and every heart beat is a gift of God. The cessation of either has fatal AND eternal consequences.

Theses Sixty Seven to Seventy Three: Fossils and Radiometric Dating of Fossils

Premise: Fossils show NO Evolution taking place AND sudden appearance in the fossil record; therefore there is no valid reason to assume a gradual, evolutionary sequence. Radiometric dating errors of fossils and other finds show that this manner of dating is NOT reliable in this application and therefore DOES NOT 'prove' great age of these fossils.

Theses 67: ALL major animal types appear suddenly and without transitional forms in the fossil record.

Evolutionists have recognized this and fully admitted this is the case. Even their hero Charlie D indicated that if his theory were true that as the fossil record was more completely examined, multitudes of intermediary fossil types would be found.

To date, no true and undisputed transitional fossil of any type has survived independent examination. ALL - every last one - has been reclassified or shown to be a FRAUD.

In preparing this book I often looked for Evolution friendly sites by Googling certain questions and seeing what came up. For nearly every question, MOST of the sites that pop up are Evolution supporting. When you Google, 'Are there any undisputed transitional fossils?' - in the first ten listings you find only one site that claims such fossils exist. Go to the site, however, and you find the standard list of DISPUTED fossils found on the other sites. Therefore, these fossils are ONLY undisputed in the mind of that site's owner.

Fossils come in two main forms - replacement and impression. Replacement fossils are those where minerals have replaced the original bone and / or skin, etc. Impression fossils are things like leaf prints where the leaf has decayed away or foot prints.

According to the Evolution Religion - the mere existence of fossils is proof positive of their 'millions of years' dogma. They claim that it is obvious that 'fossil making' is a very slow process where bone is replaced by minerals. Yet they ignore a few things or dismiss them outright.

There are finds of dinosaur bones that indicate the bone is NOT fully fossilized! Inside the T-rex dubbed Sue was found SOFT TISSUE that was determined to be blood vessels. The tissue was later examined and verified to be from the dinosaur, not something else that replaced the original material.

Hmmm…. just exactly how does SOFT TISSUE survive for a supposed 65 million year stint in the ground? I would call it a miracle that it has survived 4400 years since Noah's Flood!

The Evolution Religion high council was instantly thrown into a tizzy and called an emergency meeting! - as this find was considered so highly improbable that their 'must be protected at all costs' timeline of hundreds of millions of years was in danger of openly being rejected by the average Joe. They could maintain their charade if only the Creationists did not believe them - but their house of cards falls once the average man figures out it is all a scam.

Threatened with expulsion and censure from the Ruling Junta; the team that found Sue's soft tissue quickly did 'some more tests' - desperately trying to find ANY way the soft tissue could have lasted a very long time. Using iron rich blood, they have been able - UNDER LABORATORY CONDITIONS - to preserve the fragile blood vessels of modern ostriches for two whole years! Why ostriches? - well, because modern birds evolved from dinosaurs don't you know?

Hmmmm…. laboratory conditions and a different critter's blood vessels….. preserved successfully for TWO WHOLE YEARS. This obviously can be hyper-extrapolated to mean that blood vessels NOT kept under laboratory conditions from a DIFFERENT critter will survive for 65 million years. Give me a break!

Folks, do you yet understand the type of people you may have been believing? Did they tell you ALL the details or just their tortured, contorted conclusions? Does knowing the details change your level of belief in what they have claimed? It should.

Did you know that the same article said about HALF of dinosaur fossils now re-examined have soft tissue still in them and that this soft tissue has survived up to 199.6 million years? Are you kidding me?

You are welcome to keep on believing them if you want to - but DO NOT fool yourself into thinking you are believing science, because you are not. If you continue to believe their timeline, you do so IN SPITE of the evidence - not because of it.

Theses 68: Fossils of modern, still living critters are found in many places around the world.

Probably the most famous face-plant for Evolutionists was when the Coelacanth (lobe finned fish) was found still alive in 1938 off the coasts of southern Africa. This fish was found to be a rather common catch by the locals who ate it themselves or sold it for food in the fish markets. It had been found in the fossil record and declared to have gone extinct some 70 million years ago. It is still considered to be THE index fossil for the '70 million years ago' layer.

If a lobe finned fish fossil is found, the Evolutionist instantly assigns an age of 70 million years ago to that layer. What if an OOPART such as an three-pronged, stainless steel spear head was found stuck into the fossil fish? I have known Evolutionists so wedded to their dogma who would IGNORE the obviously man made OOPART and cling to the timeline index fossil! For the record, no one has found such an OOPART, I was simply trying to indicate the level of belief some Evolutionists I have argued with have demonstrated.

If there was no one else around and the Evolutionist chose to - he could simply dispose of this evidence and not report the find. This is WHY all the fraudulent claims and mistakes of Evolutionists DO matter.

If only one deception had happened.... a reasonable person might say this was a single bad actor - but that is NOT the case here at all.

Piltdown Man - FRAUD. Ernst Haeckel FAKED embryo drawings. MISTAKEN identity of Nebraska Man (found to be a pig). Saying Coelacanth was extinct then finding it alive. The list of errors and deceit is long!

Normally guilt by association is not a good practice to engage in. However, if a group has as bad a track record as Evolutionists have, the CREDIBILITY of ALL Evolutionists is STUNNINGLY damaged - to the point where a person is perfectly justified in refusing to believe anything they put forward UNTIL it has been thoroughly examined by Creationists and NO other better explanation is determined to exist. By this criteria - there is NOT EVEN ONE single piece of evidence that supports Evolution.

Consider the 'fossil tree' that supposedly died out many millions of years ago - until it was found alive in southeast Asia and is now being cultivated! There are many examples of insects and shellfish thought to have been long extinct but found alive today.

This begs another question: How could the Coelacanth have NOT CHANGED at all for 70 million years while man supposedly went from a lemur-like critter to what we are today in just 4 million years? Why has the praying mantis and scorpion found in fossil amber not changed at all in their imagined 150 million years while man has changed so much?

The answer is NOT some convoluted, unprovable claim that these creatures have attained their highest possible form and entered 'stasis'. The CORRECT answer is that Evolution is a FALSE RELIGION that demands illogical, fantastic, unprovable explanations every where we turn! If Evolution is WRONG - which it is - then there is ONLY ONE valid contender left standing. That is Creation.

Creation thought matches everything it is logically applied to with no mental gymnastics required. Every where we look we CAN explain what we see if we will simply allow for a Creator God to be considered as a possibility.

Your choice - the impossibly magical pretzel 'logic' of Evolution or the easy to accept, completely logical concepts of Creation.

Theses 69: We have already covered some Theses regarding the use of Radiometric Dating and the assumptions it MUST use. In these next few Theses we will look at some well known contradictions of this pillar of the Evolution Religion's timeline.

Until radiometric dating was conjectured to be able to be used to verify the old ages demanded by Evolution, all they had was the circular reasoning of the imagined 'Geologic Column' with its 'Index Fossils'.

You see, the Geologic Column purported to know the age of certain layers represented by Index Fossils. First they claimed a layer was a certain number of millions of years old and then assigned an Index Fossil to that layer.

From that point on, whenever that fossil was found - it was automatically assumed the age of the layer was known. Classic circular reasoning - where one assumption is counted as proof of another assumption. This is actually just smoke and mirrors to sound scientific when it is not.

Creationists called the Evolutionist out on this obvious circular reasoning scam - and so, in an attempt to give more credence to the method - the Evolutionists began using radiometric dating as an 'independent cross-check' on their method.

However, as pointed out earlier - at its foundation; radiometric dating relies on unavoidable ASSUMPTIONS. Therefore it should come as no surprise that this method yields unusual and impossible dates for the Evolution camp on a routine basis.

What do the 'eminently qualified and completely unbiased' scientists (gack!) of Evolution do when the measured date from the lab does not match their prediction? Why they just say the sample was contaminated and ignore the result and keep submitting samples UNTIL they get something that matches!

This is EXACTLY BACKWARD from what a real, honest scientist would do. A real scientist would ADJUST HIS THEORY to FIT THE DATA, not ignore evidence that disagrees with his theory. This tendency to ignore contrary but valid evidence PROVES that Evolutionist 'scientists' are nothing but promoters of their belief system - their RELIGION.

Theses 70: Leaching out affects the radiometric dating 'age' of a sample of rock. Chemicals compounds and elements exposed to the environment exhibit both leaching (removal of by migration of the element out of the sample) as well as some cases of concentration of a chemical if it soaks a sample and migrates into the small spaces between the molecules of the rock sample. Far more likely is the leaching effect - but in EITHER CASE - the base assumption of radiometric dating is affected and will cause a result that is NOT ACCURATE.

If leaching out of the radioactive element in a rock sample has occurred - there will be an erroneously SMALLER amount of the radioactive isotope remaining in the rock. This will show up during a radiometric dating attempt as the rock being far older than it really is.

I worked at an American site called the Defense Waste Processing Facility. Basically we took left over radioactive and chemical waste from the era of America making nuclear bombs and processed it so that the radioactive part was trapped in the crystalline structure of a glass matrix. We did this to MINIMIZE the amount of radioactive isotope that could ever leak out before it decayed into a non-radioactive isotope.

Notice I said we did this to MINIMIZE the leaching out of the radioactive element. Even with this highly complex process we could never completely STOP the leaching out of the isotope. Because this was so, we poured the molten glass - now containing the radioactive isotope - into huge stainless steel containers so it could cool and solidify. Once cooled, even if some leaching out did occur - it would be captured in a permanently welded shut stainless steel canister.

The canister was then deposited into a shielded vault until such time as it could be sent to the federal repository at Yucca Mountain, Nevada.

We then DOCUMENTED the date this process was completed and how much radioactivity we had put into the can.

Now just imagine that decades later AFTER the canister was delivered to the Federal repository - someone decided to open a canister and remove a sample of the glass and measure the amount of radioactivity still in the glass.

We KNOW that SOME isotope would have leached out, but not how much. We KNOW the laboratory measuring the sample MUST assume how much radioactivity was put into the can in the first place - unless the ones that put it there produce the records attesting to how much had been put in.

If we did not tell the laboratory how much radioactivity was put in the can - they would have to guess. Maybe they would guess right, maybe not - maybe they would be way off! IN NO CASE could that laboratory KNOW for sure how old that sample actually was based only on the amount of radioactivity they measured. They could GUESS - but they could NOT know.

What if they were only provided the sample and not even told it came out of a can? Now all they have is a rock (glass) sample that looks like obsidian produced from a volcano. If the lab personnel believe in Evolution's billions of years; they may well date the sample to match what they think is true about the sample - even though the sample is actually only a few decades old.

The bottom line is this: Radiometric dating is an inexact way to TRY to determine age.

One can find on-line passionate defenses of the methodology and also find in the same search results those that show the claims made in the defense are unfounded. Until such time as there is a reasonable answer given that cannot be refuted by others competent on the topic - radiometric dating cannot be trusted as the sole arbiter for a question so important as the Bible's timeline being True or not.

I read the entire TalkOrigins defense of radiometric dating - and I find that defense wanting for the simple reason that this defense NEVER tells us how the heavy element isotopes came to be in the first place! It does not address the missing element Technetium and just assumes that since we have all these 'heavy elements' now that they have no obligation to tell us how they came to be.

Neither does the article adequately address the Polonium radio-halo work of Dr. Gentry. Neither does the argument tell us just exactly how electrons can remain in stable orbit for billions of years. It is a classic case of offering a possible (in their mind) explanation of SOME aspect of the overall Creation vs. Evolution argument in theoretical terms. It SEEMS to work and give consistent results for VERY limited cases - therefore we are supposed to believe absolutely everything about Evolution from all disciplines of science.

The defense is very technical in nature and makes several claims that knowing the initial concentration of isotope is unnecessary. Having worked in nuclear power for decades and WATCHED how certain isotopes are measured when we DO have a good idea of what the initial concentration SHOULD BE - I know from FIRST HAND knowledge that a very small error in setting the instrument doing the measuring can throw off the final result off by factors of ten.

The TalkOrigins defense NEVER deals with the effects of known leaching in the real world. It ONLY deals with THEORIES and METHODS claimed to work by FALLIBLE (albeit very intelligent) humans.

This leaves us with the known circular reasoning regarding the geologic column and index fossils that the prophets of the Evolution Religion tried to pass off as real science - with no clear consensus on the validity of radiometric dating's accuracy.

Theses 71: Given no clear consensus on radiometric dating - as any cursory review of the topic shows - let us look at some published results that tell us the method is likely not to be depended upon.

The first is the FACT that the fossils are NOT generally directly radiometrically dated! - because they are usually found in sedimentary rock; which the TalkOrigins defense admits is nearly impossible to date.

That's right, the anthropologist finds a 'Lucy' in sedimentary rock and then goes looking for the closest lava flow and gets a sample. Sometimes they get two samples, one from above the fossil find and one from below it - so they can get two dates.

They get the sample(s) dated and then INFER that the fossil is of an age related to the 'radiometric age' of the lava sample(s) - but they tell all of us that they KNOW EXACTLY how old the FOSSILS are. This is a

LIE, they DO NOT KNOW - they MAY have an indication of how old the lava near the find is - IF radiometric dating is actually a valid method AND they got a 'good' sample.

The TalkOrigins defense tells us how easily samples are contaminated or affected by certain factors that only the laboratory experts can spot. The laboratory analysts can ONLY take the anthropologist's word for it regarding the collecting of the sample.

The finder of the fossil will also often BIAS the analyst by telling them how old he THINKS the sample is - because there is big fame and fortune IF you find an 'older ancestor of man' than has been found so far.

If the samples do not bear out this age - or something close - the sample is ASSUMED to not be valid and this data point is discarded. It NEVER enters their minds that maybe their METHOD is flawed and/or their ASSUMPTIONS are incorrect.

The TalkOrigins defense complains about samples of lava Creationists submitted for dating that were known to be from 1802. The Creationists did NOT tell the analyst how old they thought the rocks were.

These samples measured in the hundreds of millions of years old range! The TalkOrigins article gives excuses for this utter failure of radiometric dating - claiming sample contamination, etc.

The point is - ALL of their samples that they analyze come from the REAL WORLD and NONE of them have been 'laboratory controlled' for their entire existence.

The TalkOrigins defense admits that there are several ways the samples could have their 'clock' reset; such as re-melting. The article also says contamination or poor quality samples may indicate a wide variety of ages within the SAME lava flow.

Hopefully you can at least see why Creationists raise a skeptical eyebrow and cry foul when Evolutionists finally find a lava sample that comes close to their biased expected range. They immediately go out saying they KNOW this means the fossil found nearby is EXACTLY 2.3 million years old and anyone who disagrees is some kind of reality denier. They NEVER tell the media or general public how many samples they had to throw out before they got what they were looking for.

To be sure: We ARE dealing with reality denial - just NOT on the part of the Creationists.

Theses 72: Living mussel shells were carbon dated to be 2000 years old. Again, a defense from an evolution supporting website simply dismisses any and all evidence submitted by a known Creationist out of hand.

That defense is from the National Center for Science Education - an organization founded and completely controlled by Evolutionists - but named to sound objective. The site complains that these mussels got their shell material from the surrounding limestone which was very much older and so this threw the measurement off.

Ummm…. just exactly how does this defender KNOW this is the case? Again we see a passionate CLAIM with NO supporting data -

ONLY an excuse as to why C-14 did not give a reasonable age for the LIVING critter.

The same website's Q & A defense of C-14 dating - in the very next answer to the question of C-14 being still found in many 'hundreds of millions of years old' coal samples when there should be NONE - claims the sample is too old, so the measurement of C-14 must be from contamination!

Ummmm.... the ONLY C-14 available to contaminate the sample would be from the atmosphere. Just how did it get into the BURIED COAL? Again, how can the defender KNOW this when he DID NOT do the measurements? He cannot know and does not know - but since these data points PROVE his RELIGION is FALSE he MUST do whatever he can to take them down.

Theses 73: Woolly mammoth carbon dating discrepancies.

I have taken Evolutionists to task a lot in this book and for good reason. They have been caught numerous times lying, grossly misinterpreting evidence and making highly unreasonable assumptions.

The Evolutionist side quickly piles on ANY error however slight that is made on the Creationist side - and a few errors have occurred.

One is discussed here: There HAVE been times where Creationists have claimed that different parts of MAMMOTHS and the fauna associated with them have yielded different ages by C-14 aging methods.

Some of my own heroes of the Creationist movement have even put this idea forward - apparently never checking the existing report from which the data comes to see if what they repeated is actually in the report.

A couple of evolution supporting sites found the report my friends were referring to and pointed out that the information my friends were putting forth is no where to be found in that document - at least in the way my friends characterized the information.

My own admission: In previous books; I simply took what they said and repeated it. I trusted them as for 99.9999% of everything else they said **HAS** stood the test of time and the attacks of Evolutionists. However, no one is perfect.

I should have been more thorough - and that is why, as I am writing this book, I am checking facts primarily using Evolution friendly sources who have critiqued earlier Creationist works.

This time I **DID** look the source document over myself - and found something very interesting!

While it is true that there is no mention of a SINGLE MAMMOTH having different parts of its body yielding different C-14 dates - there are SEVEN samples in Table 4 of other animals and fauna FROM THE SAME LAYER that have different carbon dates FAR OUTSIDE the accepted range of each other.

Table 4 shows all the analyzed samples from all the different layers from all the different areas discussed in the report. The table's arrangement is odd, in that some of the data you would expect to be next to each other sometimes has several lines between entries or even on another page.

I will not go over the entire 40+ page report, but will focus on ONE LAYER from ONE AREA. Look, if they screw this up badly - then I would have a very difficult time believing they got ANY of it right.

Here are seven entries from Table 4 from the same layer identified only as '...in frozen silt, rich in organic matter...'. The samples were sent to different labs - or at least analyzed at different times - with six samples giving their date of collection which ranged from 1937 to 1951. We start on page 30 of the report, several pages into the very large Table 4 with:

1) Bison hide; 11,980 ± 135 years old; lab SI-1633; collected ???

2) Hair from hind limb of Ovibos; 17,210 ± 500 years old; lab SI-454; collected 1940

3) Muscle from scalp of Ovibos; 24,140 ± 2200 years old; lab SI-455; collected 1940 (there is a note saying this is the same animal as the sample analyzed by lab SI-454 in item #2 just above)

4) Bison; 11,735 ± 130; lab SI-1631; collected in 1937

5) Tendon of left tibia of *Felix atrox.*; 26,760 ± 300 years old; lab SI-355, collected in 1938

6) Dung?; >40,000 years old; lab SI-291; collected in 1948

7) Hair from skull of mammoth; 32,700 ± 980 years old; lab SI-1632; collected in 1951

My Creationist friends said that two samples of the same mammoth had yielded two vastly different C-14 dates and I have repeated it.

This was an error - BUT:

The basic PREMISE was that using C-14 isotope measurement is unreliable BECAUSE samples from the same animal can yield vastly different dates.

Table 4 DOES show this exact problem for items 2 & 3 above - but the samples came from Ovibos not a mammoth. Ovibos and a mammoth WERE found in the same layer. Therefore, the PREMISE is valid, though the data we all referred to was somewhat in error.

The C-14 date for item 2 above was $17,210 \pm 500$ years (i.e. 16,710 to 17,710 years old). What was the carbon date of the second sample? Drum roll please.... $24,140 \pm 2200$ years (21,940 to 26,340 years old)!

If you subtract the MINIMUM age of the second sample from the MAXIMUM age of the first sample you get:

21,940 - 17,710 = 4230 years discrepancy for the SAME ANIMAL collected / analyzed in the same year of 1940 done by two separate labs. Further; this critter was found in the same layer as FIVE other critters and fauna ('dung?'). Those samples have a date range from 11,605 years old to > 40,000 years old.

My creationist friends had also charged that fauna associated with a mammoth find showed a different age than the mammoth. Again, they got the data details a bit wrong - BUT the underlying PREMISE that measuring C-14 is unreliable still stands because the fauna identified as 'dung?' in Table 4 dated > 40,000 years old while the critters that likely produced the dung have numerous dates associated with them; some being tens of thousands of years younger!

Regarding the mammoth data - to correct the record: The mammoth is nearly 5000 years older than Ovibos and about 8000 years younger than the dung! Therefore: The basic charge made by all Creationists is valid.

In summary: C-14 is marketed as a bullet-proof dating method by Evolutionists DESPERATE to 'prove' the Bible's timeline wrong. The actual data points show that it is the C-14 dating method that is wrong. If the METHOD is WRONG - then ALL the CONCLUSIONS drawn by Evolutionists using that method are also WRONG.

Further, since we Creationists are human, too: YOU must look up the base / source information YOURSELF when counter claims exist. No one can take the time to look up everything. In my books, I encourage people to do their own research and come to their own conclusions.

I apologize for only NOW following my own advice on this one point. All I can do is correct the record whenever it is shown to need correcting and I have now done this. YOU must make a determination of WHO is the most credible reporter in this debate.

In your determination, consider:

A relatively minor mistake - that did not change the basic premise - admitted to and corrected as soon as it became known on the Creationism side

Versus -

Outright fraud (more examples still to come), withholding of critical evidence, gross extrapolation of known data to draw unwarranted conclusions on multiple occasions on the Evolutionist side.

At least when a Creationist makes a mistake - generally they confess it and correct it.

Theses Seventy Four and Seventy Five: Chemistry - Salt (Sodium Chloride) as Well as Thousands of Other Predictable Reactions

Premise: There are very thick layers of almost pure sodium chloride found in many places around the world. Since the reaction of the base elements sodium and chlorine is extremely EXOTHERMIC (releases heat in the process of combining); there is no way the amount of salt we know of could have formed without devastating effects on the rest of the planet - and these effects would have left physical evidence behind.

Theses 74: Evolutionists like to claim that they 'own' Chemistry in the Creation vs. Evolution debate - but nothing could be further from the Truth. I will not go into depth here, but there is ONE phenomenon that the other side would like to simply ignore:

Just how did ALL THE SALT in the world form without leaving a trace of the heat that would have been released in such a reaction?

Understand that these salt layers can be THOUSANDS OF FEET thick according to mining websites. Generally it is claimed on Evolution supporting websites that this salt is the result of water evaporating and leaving the salt behind.

There are a couple of really obvious problems with this line of 'thought':

1) It does not tell us where the ORIGINAL SALT came from that was dissolved in the sea water - therefore does NOT address the exothermic heat issue at all.

2) These layers are nearly pure sodium chloride - so much so that often NO purification is required before simply packaging it for human consumption - if evaporation is how it came to be; why not lots of inclusions like dirt and organic matter mixed in?

3) Recall that the oceans today are slowly getting saltier. Evolutionists would have us believe that vast oceans repeatedly evaporated - yet today we see the Dead Sea and the Great Salt Lake slowly losing water level but NOT leaving behind anywhere near the pure thick layers we find UNDER hundreds of feet of rock and dirt.

4) Recall from an earlier Theses that Evolution's spontaneously generated first super cell supposedly evolved in a warm pond - yet it was shown that even a little salt in solution PREVENTS the DNA from attaining the very small size we find in every cell today.

Worse than the exothermic problem for Evolutionists is the fact that BOTH sodium metal and chlorine gas in their elemental state are LETHAL to life! Think of the vast quantities of free sodium (a solid metal) and free chlorine (a gas) that you would need to make all the salt we find!

Also, exactly HOW do you get all that METAL sodium to react so completely with ONLY the GAS chlorine when "Sodium is highly reactive, forming a wide variety of compounds with nearly all inorganic and organic anions." - according to brittanica.com?

When did this event take place in the Evolutionist timeline? You will search their scenario in vain for this event. It is simply ignored because it presents still more impossibilities for their story to have any chance of being true.

Creator God knew that the humans He created were made needing salt in a non-lethal form - so He made it for us. He also made a LOT more of it because we would need it for other chemicals and uses, but most importantly as yet another attempt to reach out to fallen man who want to say He doesn't exist.

This is WHY we find so many unexplainable conundrums for the Evolution Religion. Father God TRULY DOES want ALL MEN - even atheists and evolutionists - to come to a saving knowledge of His Only Son Jesus the Christ.

Theses 74: Above I shared a single example of something the other side cannot explain and does not even attempt to. The field of chemistry holds many thousands of 'equations' that represent known, predictable chemical reactions. Man discovers many more every year.

To understand and predict a reaction - we have to understand the atomic model of unseen electron shells. We can then predict if two elements will react. These are 'Laws of Chemistry' if you will. Where did these 'Laws' come from?

The Evolutionist MUST say this VERY ORDERLY, predictable process of assembling (an intelligent act when done in a laboratory) is a result of PURE CHANCE! As stated so many times before - you are welcome to BELIEVE them, but you do so by PLACING YOUR FAITH in the 'intelligence' of an admittedly flawed human being.

Yet, if you choose to BELIEVE that Creator God is the Ultimate Chemist able to create the elements AND put into place the Laws that govern their behavior - you ALSO place your faith in Him.

You MUST place your faith in one of the two propositions. Just because it is EASIER to believe in a 'God did it' scenario DOES NOT automatically make it the wrong view!

The Creationist DOES NOT simply throw up their hands and wimp out by believing in Creator God. They simply look at what man knows about each subject (in this case chemistry) and draws a LOGICAL CONCLUSION - and then acts accordingly. The Evolutionist does exactly the same thing - but draws a different conclusion.

That some college professor personally does not believe in God DOES NOT make his view any more valid! It also DOES NOT justify his failing or harassing a Creationist student because he does believe in God. The Evolutionist may have come up with a snappy label for when the Creationist simply admits that God is the BEST answer - saying that we are appealing to 'the God of the gaps'. This may be good marketing, but DOES NOT automatically make the Creationist idea wrong.

Evolutionists have NO IDEA where these Laws came from and yet they DEMAND that everyone submit to their mantra, "It all happened by CHANCE." Are they not appealing then to 'the god of chance'? Either way you choose - the path is 100% RELIGIOUS - for it depends solely on what you BELIEVE.

Theses Seventy Five:
Biology: Bio = LIFE; ology = STUDY OF

Premise: Since the origin of life has NOT been proved possible by abiogenesis - it is perfectly acceptable to invoke Creator God as an explanation.

Theses 75: Life is COMPLEX at EVERY LEVEL we look at it! Cells are extremely complex as are the chemical building blocks of the structures of the cell or cells. We see flagellum motors where EVERY SINGLE PIECE of said motor MUST have come into existence at the same instant for it to have had any functionality at all.

This concept has a name: Irreducible Complexity. It is covered very completely in the book Darwin's Black Box by Michael Behe. The essential argument goes like this: If I see complexity that requires intelligence to understand - then it required intelligence to bring it into existence. Creationists believe in an Omnipotent (ALL-Mighty) and Omniscient (ALL-Knowing) God - He is not only the Ultimate Chemist, but is also the Ultimate Biologist, Geologist, Everything-ologist!

I demonstrate the idea of Irreducible Complexity in my talks using a basic mousetrap. First, I show the audience a fully assembled trap. It has five pieces: Mounting Board - Spring - Trap Bar - Bait Bar - Bait Bar Holding Bar - (and some strategically placed staples to hold various parts together). I then show the crowd a series of traps that have ONE of the five main pieces removed. It is immediately obvious that NONE of these traps will do more than feed the mouse! ALL FIVE pieces are required to be present AND installed in exactly the correct way.

However, to catch a mouse - the trap must ALSO be baited with the correct type of bait and set using the holding bar. Many a person has ended up snapping their own fingers trying to set that hair-trigger bar! I know that I have.

There is still more! This device is irreducibly complex in that no main part can be missing and it still work - but remember that the device WAS DESIGNED to catch something - a MOUSE!

ANY TIME you see something DESIGNED with a definite purpose; it is perfectly LOGICAL to INFER that there was a DESIGNER.

I most certainly have NEVER MET the designer of the original mousetrap - yet I can be SURE that there WAS a designer of that device with a purpose. Things do not make themselves and even if they did, they would have no set purpose.

What is the Evolutionist's explanation of complexity and purpose?

Why - all that is just the APPEARANCE of complexity! In reality it all came about by chance with no purpose whatsoever. TRUST ME! I am the expert!!

Yes, folks - that REALLY IS their ONLY answer to this vexing issue for those who would have you believe there is no Creator.

It really DOES take a lot more faith to believe their fantastic claims than it does to simply go where the valid evidence takes you: That there IS a Creator God.

Theses Seventy Six:
Chicken or the Egg First Scenarios

Premise: If we look in nature and see an unexplainable thing by Evolutionary 'processes' - then it is logical to consider the Creation explanation as a better view.

Theses 76: Literally the ultimate party question is: Which came first - the chicken or the egg? Your answer actually reveals whether you are a Creationist (even if you do not know you are one) or an Evolutionist!

You see, if you say the chicken came first - you are a Creationist - for you conclude that God made chickens with the pre-programmed information in their complex bodies to lay eggs which hatch into more chickens.

However, if you are an Evolutionist - you CANNOT EVER admit God was involved in the creation of anything. Therefore you come up with the ULTIMATE EGG coming first. What egg would that be? Why - the COSMIC EGG of course (which was laid by the Cosmic Chicken)!

Their story goes something like this….

"You see boys and girls: First there was absolutely nothing - then, this nothing-ness burped and the cosmic egg appeared - which exploded for no reason whatsoever - and…. and…. that formed galaxies and solar systems and planets - and on Earth it rained on the rocks for millions of years and formed Cream of Pre-Biotic Soup - which… ummm…. ummm….. (STOP LAUGHING!)…. was struck by lightening… and…. and…. umm… spontaneously formed the first super amoebae…. which morphed over more millions of years into bananas and baboons… and…. (Johnny - I told you if you ever said 'God did it' again I would send you to the Principal's Office)…. and the baboon became a pre-human and …. here you are. Test will be Tuesday."

Folks - Evolution really is this ridiculous and impossible. I have long said that the domination of Evolution being taught as fact in nearly every 'school' (actually, indoctrination center) in America is the best proof that there is a master deceiver called satan.

The people who hold most professor positions are actually very smart people. How such smart people can get sucked into this LIE and so passionately defend it that they spend many hours trolling the internet looking for guys like me to argue with is simply amazing.

WHY would satan care so much about the Evolution Religion that he MUST DEFEND IT to the bitter end?

Think about it this way:

In the Book of Isaiah it says of satan that 'iniquity' was found in him while he was still Lucifer. The Hebrew word rendered iniquity means, '…to twist the Truth maliciously…'. What Truth would satan have known? He would have been told by God after being created by God - that he (satan) was a created being - and that God was a Self-Existent, not-created Being.

Isaiah goes on: Satan said that he would rise above God - i.e. dethrone God and take over His Kingdom.

This would mean he RECOGNIZED that he was INFERIOR to God when he said this. Therefore he would have to change… over… over…. TIME into a being first equal with God and then into a being greater than God.

Hmmm… change over time….

Isn't this the BASIC PREMISE of Evolution? Doesn't this concept claim that INFERIOR things can change over TIME into things more and more complex….

You see, for satan to have ANY CHANCE at all of ever realizing his VAIN ambition - he MUST first conceptualize that Creator God is not God at all - but had a beginning like all other created beings have. Therefore God could not have created anything, but…. satan then had to come up with some no-Creation explanation for ALL that he could see existed.

Hmmm…. What if God was merely a created creature a few steps ahead of him (satan), well! - he (satan) could 'evolve' faster than God and ultimately catch up to God and…. and… one day…. find himself: Greater than God! He could take over the Title of God! He would be God!

Evolution is really the SAME OLD deception presented to Adam and Eve in the Garden by nachash (the serpent - the HISSING ONE in Hebrew).

The man and woman fell to the temptation and ever since there has been war in Heaven (until Michael defeated satan there) - and then war on the Earth as satan continues to spew his **RELIGION** of Evolution to those who also want to be their own God.

If **YOU** believe something said by some being, human or otherwise - then you believe not only the idea put forth but more importantly… you believe the **PERSON** or being or Being putting it forth.

Believing carries with it a **RELATIONSHIP**.

That a real relationship exists is proven by a person defending those he shares the belief with.

If you believe in Evolution, you believe those putting forth the idea - and the being that put forth the idea in the first place - i.e. ultimately; you believe satan's lie over Creator God's His-Story recorded in His Word.

If you go further and **DEFEND** Evolution, you have a much more solid relationship with the one you have believed - and that relationship **IS AS** a follower of satan….

I know these are stern words - but consider what is at stake: **ETERNITY**.

Not to make fun of such a serious topic - but the bumper sticker says it very well:

> Eternity: Smoking or Non-Smoking - The Choice is Yours

Believe Creator God today and enter into a real relationship with His Son Jesus Christ. He will break your 'mind-chains' that satan has shackled you with.

Theses Seventy Seven thru Seventy Nine: Extinction is Observed, NEVER a NEW Organism Arising / The REAL Effects of Mutations / Barriers to Interbreeding

Premise: If Evolution were the Truth; we should see the 'arising' of totally new critters better suited to the changing Earth conditions all the time - but we NEVER do, EVER! Therefore, MACRO-Evolution (a.k.a. VERTICAL Evolution) remains just a CLAIM of Evolutionists after some 140 years.

Theses 77: It is a known fact that interbreeding of critters is not possible above the Family level classification. If the animals are below the Family designation, they MAY be able to interbreed and this MAY result in some interesting 'new' critters - but the resulting animal WILL BE the same basic kind of animal - always.

Evolutionists claim that because we DO SEE very minor changes in a critter population due to some change in its environment - that all these minor changes ADD UP over time to create TOTALLY different critters from its parents. They CLAIM that it happened in the unobserved past and is happening today - but is happening so slowly that a man's short lifetime will never record it.

If the baby critter inherits a mutation - it may have an extra leg or two heads - but still remains basically the same critter. We will discuss mutations more as Theses 78.

What DO we in fact see? We DO see species go extinct - which brings no one any pleasure - but we have never, ever recorded a totally new creature coming from two parents that are able to breed with each other. We have NEVER seen a frog reproduce anything other than more frogs.

We mate a donkey with a zebra - both critters BELOW the Family level - and they have a zee-donk baby that has characteristics of both parents - but it is STILL basically the same animal.

We zap fruit flies with massive doses of radiation or chemicals and their babies have curved wings or no wings - but the baby is STILL just a FLY (well, in the case of the fly with no wings we might call it a 'walk').

Theses 78: One way to get a 'slightly different' off-spring is by mating very similar basic animal kinds - but this has shown to ONLY result in a basically SAME kind of animal. Dead end for Evolutionists. What mechanism will they promote next? Ummm…..

Mutations!

Professor Noah Itall states confidently to his class, "Yes, yes! Mutations happened that offered some advantage to the baby in the struggle for life!" {Gotta give it a snappy name…. ummm….} "Beneficial Mutations! YES! That's it!" {Now I must put on my serious scientist face and tell everyone that a pair of dinosaurs with only scales mate and out of some of the eggs hatches a baby dinosaur with FEATHERS - or pre-feathers at least!} "Yes, yes! This is really believable…" {SO WHAT if I have no fossil evidence to back it up}. "It is a fact! How do I know? Well, we have dinosaur fossils and we have birds today - plus we KNOW Evolution is a fact and they look {to me} like they are related - therefore since I am the self proclaimed expert in all things; it is an indisputable FACT that dinosaurs evolved feathers and turned into birds!" {Case closed. Oh, crud! That pesky Creationist kid has raised his hand.}

"Ummm……. professor Noah Itall?"

"Yes, Johnny."

"Our biology science book says that mutations are almost never 'beneficial'. It says they mostly do not change anything or they hurt the critter but it recovers but does not pass on the mutation or the mutation hurts the critter and it dies or the mutation makes the critter sterile so they could never pass the benefit on. Can you explain this?"

"Johnny - GO TO THE PRINCIPAL'S OFFICE IMMEDIATELY! The rest of the class, exam will be on Tuesday - and I expect everyone to have memorized these facts I have stated them or YOU WILL FAIL! Class dismissed."

Folks. Even so-called 'beneficial mutations' are the result of a LOSS of information - NEVER a GAIN of information. There is NO WAY a fish can mutate fins into legs. The lobe finned FISH Coelacanth is STILL JUST A FISH - and the lobe fins are TOTALLY useless for walking on land. Even if these lobes could support the FISH out of water - it would DIE because it cannot breathe air.

Also, the simple NUMBER of information gaining mutations necessary to turn a bare skinned arm of a dinosaur into a useful wing numbers in the tens of thousands (at least). In the meantime, the new critter would have non-functional arms AND useless wings! How is this situation an advantage? This critter would be far more likely to DIE.

Still another problem! What critter of the opposite sex is this new critter going to marry and have kids with? THAT critter would have to evolve the exact SAME mutation at the SAME time in the SAME place!

Please tell me that you have realized by now just how much the Evolutionist DEMANDS that you swallow their story whole based on his UNPROVED claims!

YOU first must get un-brainwashed and then will have to un-brainwash your kids if you send them to public school or a private school that ascribes to the Evolution Religion.

Theses 79: A cat cannot interbreed with a dog to give us a Cog or a Dat! While we can observe SOME very limited interbreeding of critters that LOOK different, but are able to bear off-spring as in the case of the Donkey and Zebra having the odd-looking Zeedonk - all this shows is that these parent creatures MUST be closer related than we were previously aware.

Dogs today come in hundreds of 'breeds' - and thousands of 'mutts'. Still, they are ALL dogs. Where did they come from?

Creationists would claim that they are ALL off-spring from ONE original pair of a 'dog-kind' animal that was created by God and rode on Noah's Ark through the Flood and was released. Humans tamed some of their kids and bred the various types of dogs we see today. In the wild, the same kind of thing happened to produce foxes and wolves and coyotes.

More than one college professor has laughed a Creationist to scorn for this claim. However, let us look at what this Evolutionist professor is teaching his class. The dog had a 'common ancestor' with all other life forms - i.e. that spontaneously generated super amoebae! It 'arose' - whatever that means - from DEAD CHEMICALS after it had 'rained on the rocks for millions of years'.

At least the Creationist says dogs give rise to other types of dogs!

The Evolutionist says the great-great-great...... great grandpa of your teacup poodle was a ROCK!

The Creationist's view AGREES with the KNOWN Laws and principles of reproduction. The Evolutionist's view VIOLATES the Law of Biogenesis AND many known principles of reproduction.

Remember: Survival of the Fittest DOES NOT EXPLAIN Arrival of the Fittest!

Theses Eighty thru Eighty Four: More Conundrums Evolution Cannot Explain BUT Creation Can!

Premise: If there are valid examples of mutual exclusivity found in nature that can only be explained by ONE of the two major Theories of Origins - then the Theory that CAN explain the conundrum IS the CORRECT concept.

Theses 80: Like the previous 'which came first, the chicken or the egg' question; the next few examples found in nature cannot be explained at all by Evolution. If we attempt to understand these conundrums from the OPPOSITE viewpoint to Evolution - which is literal, young Earth Creation - we see these phenomena of nature nicely explained and actually predicted by the Theory of Creation.

Some creatures can only survive by eating certain foods. According to the website MindFloss; among them is the Monarch Butterfly. In its caterpillar stage, it only eats the leaves of the milkweed plant. The same website indicates that the pen-tailed tree shrew of Thailand ONLY drinks the naturally fermented nectar of the Bertam Palm.

Think about that for a minute. If the Theory of Evolution of molecules to moles to man were actually true - then BOTH the critter AND its exclusive food would had to have evolved in the same place at the same time for no reason whatsoever by pure chance. Sorry, did not happen.

The Venus Fly Trap is a carnivorous plant - it has the ability to catch small insects and digest them and repeat the process.

Just exactly HOW did this plant KNOW how to change itself from a non-insect eating plant into one that eats insects? HOW did it KNOW that it had to emit a sticky substance and develop a triggering mechanism so that its trap would close around the victim bug? If this NEW way of nourishing itself needed a million years to evolve - HOW did the plant KNOW that there would still be insects to catch in a million years?

The answer is easy folks! The Venus Fly Trap DID NOT evolve these characteristics!

Let us apply Creation Theory to these issues. The Creator DESIRED to Create a variety of creatures and cool plants - and so He DID! Does this claim MATCH the LOGICAL evidence - OR - does the unbelievable contrivances of Evolution fit better?

Just because the Creation concept is SIMPLE enough for a child to understand and accept DOES NOT automatically mean it is wrong! Remember that Jesus said we MUST have FAITH to be saved! Further, He said that faith had to be like that of a little child.

Would it not make sense then that His Father would Create a world filled with wonders so unexplainable that humans would conclude there MUST BE a Creator? If that Creator at the same time WANTED to HUMBLE the 'wisdom of the wise' - those who are smart in their own eyes - would He not Create some things that totally confound any attempts to explain it without God?

Moreover, if this Creator God wanted His created humans to give Him praise for His Creation - not because He forces them to, but because they recognize what it would take to do all that they observe and understand - would He not make the very unexplainable things we see?

Yes! - to ALL THREE statements / questions!!

Theses 81: Symbiotic relationships are another unexplainable aspect of the Creation that point directly to a Creator! Obligate Symbiosis is defined by study.com as when two organisms in a symbiotic relationship cannot survive without each other.

As with the problem of an organism requiring a single food - this type of symbiotic relationship is IMPOSSIBLE for Evolutionists to explain without invoking the most fantastic of tales!

I looked at several Evolutionist websites regarding this topic - one claimed that 150 million years ago; one of the two organisms (that are now in an obligate symbiosis relationship) ATE the other organism - but did NOT digest it! Of course those making this claim WERE NOT THERE to observe this take place.

Indeed; man was AT this stage of evolutionary development 150 million years ago!!! The other evolutionary sites simply parrot the 'party-line' claim; 'it happened because we see obligate symbiosis today (and there cannot be a God to attribute this to)'.

If we simply allow for a Creator to be involved - KNOWING that He needed to have SOME of His creatures in an obligate symbiotic relationship SO THAT a critical part of His world would provide something CRITICAL for His most special created beings - that would be man and woman - the problem is a non-issue!

These God-deniers do not deny Creator God because the evidence is not clear; they deny Creator God because they WANT TO BE God in His place.

Theses 82: Unexplainable transfers of information. Again we go to the Monarch butterfly and similarly to the salmon. BOTH creatures migrate thousands of miles and each generation DIES BEFORE the next generation hatches from its eggs. Yet each creature - by the MILLIONS - accurately 'knows the way' it should go to successfully repeat the life cycle.

Man can transfer programmed information by INTELLIGENT MEANS using thumb-drives and email; but notice - this REQUIRES intelligent acts using intelligently designed equipment and networks powered by INTELLIGENTLY designed and operated power plants and ASSUMING the information will be put to use by an INTELLIGENT being on the receiving end.

That is an awful lot of intelligent action going on AND it is DWARFED by the ability of the Monarch and Salmon to lay an egg WITH the PRE-PROGRAMMED information ALREADY resident!

Let us recap this: MAN cannot transfer information without a lot of intelligent intervention - BUT Evolutionists would have us believe that BUGS and FISH can transfer information by a mechanism operating on sheer chance for no reason whatsoever. Tossing the BS flag now!

Folks - there is a God and He Created all these wonders so that YOU would give up your vain attempt to be Him and regain your status as one of His children by putting FAITH in His One and Only Son: The Lord Jesus Christ.

Do it now if you have not already or are not sure if you have.

Theses 83: DNA and RNA. This is probably the MOST complex 'which came first' problem in the world.

You see, DNA cannot come into existence without RNA to make it - yet RNA cannot exist without DNA....

Yup. Creator God ONCE AGAIN knew that man would one day discover DNA and RNA. Man would figure out that there is no way for one to exist without the other fully functional. He foreknew that this FACT would confront Atheist researchers head on.

He did it this way to reach out in love to these people - so they, too, would finally give up trying to be Him in their lives.

Some actually do finally admit they were wrong and that there is a Creator God. Sadly, others do not - and make up even wilder tales about how DNA could come to be without RNA. Such persons pile on the scientific jargon and invoke the magic of millions of years - and then simply CLAIM it happened.

Everything in Evolution is built on a tottering tower of SAND called unproved and unprovable claims. One can ONLY accept these claims or not. There is not one single shred of actual scientific (observable and repeatable) evidence to show how DNA came about without Creator God.

Theses 84: Care for weaker individuals rather than just letting them die. Man (female and male man) has a very strange activity that he engages in. Man OFTEN puts self at risk of injury and even death for no other reason than to help a weaker individual of the population survive.

If 'Survival of the Fittest' is the mantra that got us here from our super-amoebae ancestor; WHY would this tendency to care for the weak be so strong in humans? Why would we PRAISE and HONOR those that demonstrate this characteristic?

If the premier doctrine of 'Only the Strong Survive' to pass on their genes - then man logically would simply let the weak or disadvantaged die off if not actually involve themselves in killing them off!

Oh, wait…. Evolutionists DO involve themselves in killing off the ones they perceive as inferior in concentration camps and abortion mills and through euthanasia and by sterilizing many via eugenics programs.

It appears it DOES matter what you choose to believe about Evolution. What you believe DOES affect what you choose to DO or at least it may keep you from putting a stop to some barbaric practice like abortion.

Think about it! In the abortion 'clinic' - the poor, scared girl is ASSURED that what she is killing is NOT a baby, but it is at the Evolutionary stage of a fish or a pig, etc.

However, recently we found out via hidden camera videos that once the human baby is dead - the abortion facility then dissects the infant and SELLS its HUMAN BODY PARTS for profit so they can afford an expensive car!

The Bible tells us that a person who WILL NOT recognize God as Creator - therefore Maker of every life - is DESTINED to develop a REPROBATE MIND. Hmmmm…. Evolutionists deny God as Creator….. then engage in and protect those that commit MASS MURDER of BABIES through abortion…. then SELL body parts…. then JOKE openly about it all!! Yup. I would say the Bible is right on this analysis.

Folks! More than 50 MILLION babies have been killed in the United States alone since 1973! What a coincidence that the monopoly teaching during all this time in public schools was GODLESS Evolution.

With the invention and widespread use of advanced ultrasound equipment - there is no denying the fetal heartbeat within weeks of conception or that what is inside a woman can feel pain or that this is a child.

Many women interviewed in unscripted, live-on-the-street interviews at the recent Women's March ADMITTED openly that abortion is murder - and STILL they choose the woman's 'right' to rule (be God over) their own body than to bring the unborn child to birth and adopt it out if need be.

Could this holocaust have occurred if EVERY child in this sin sick world had been taught that God MADE each 'YOU' special with a PURPOSE for their life? No. The Bible had to be removed as the foundation of this (and any) nation first.

So it was removed.... by replacing it with the Evolutionist's bible: On The Origin Of Species.... and their mantra of (what they claim is) Science over Religion.

By the way, not all religion had to go.... only true Biblical Christianity. Satan does NOT care what you believe in.... just as long as you do NOT believe in the Creator God of the Bible and His Only Son, the Lord Jesus Christ.

May God have mercy on our souls!

Theses Eighty Five thru Eighty Nine: Outright LIES and Misinterpretations of Evolutionists Presented as Fact Though DISPROVED Decades Ago

Premise: In science, if a hypothesis or theory - or some supporting evidence of a theory is shown to be wrong (or worse - intentionally faked); then that theory is WRONG and / or the evidence in question MUST be removed from textbooks and the general public honestly informed of the error.

In the case of the Theory of Evolution - which has had so many pieces of it disproved or uncovered as lies - and lies built on other lies; the use of the word Theory to identify Evolution is no longer appropriate, even as a gesture of generosity in debates.

The word theory in the context of Evolution implies that there is a great deal of - or at least some - true science behind it. Nothing could be further from the Truth as this book has clearly laid out.

If there is NO actual science behind the Theory of Evolution, what then is it if not a theory? All one has to do is consult a dictionary to find out.

Webster's Dictionary identifies a religion as "…any system of beliefs held to with ardor and faith…." This is not the first definition of the word religion, but usually the 3rd in most dictionaries I have consulted.

This definition of religion PERFECTLY describes the current state of Evolutionary 'thought'.

Since the acolytes of the Theory of Evolution like to spell theory with a capital T to emphasize it's importance - I spell Evolution's true identity with a capital letter as well - the Religion of Evolution.

Fully 84 theses have been briefly examined thus far, each clearly showing the absolute bankruptcy of logic and complete absence of real science underpinning the vaunted Theory (Religion) of Evolution. The Religion of Evolution has no more basis in science, logic or common sense than belief in the long discredited gods of Egypt or the pantheon of gods of Greece and Rome. It IS simply another FALSE religion that happens to exist in our day.

Just as ALL false religions eventually grow up and reveal their fruit so that all can see it for what it is: Evolution's Reign of Terror will someday fall. Perhaps it will be this book that finally bursts the dam and washes it into oblivion. If it is, again - this book would not have been had it not been for the giants that have gone before me.

A few more common nails to drive in the coffin of the False Religion of Evolution before the final ones that show Evolutionists revealing that they do not really believe their own babbling.

Theses 85: Birds are widely claimed to have 'arisen' from dinosaurs these days. The problem is that bird fossils have been found in layers LOWER than their supposed dinosaur predecessors. Do the Evolutionists who put forth the 'bird from dinosaur' claim know this? Well, if they did not know - that is a case of simple ignorance. That is bad enough; for these folks hold themselves out to be the experts.

What if they DO KNOW about the fossils - but make the claim ANYWAY. Ah…. now they are LYING! Now they are PURPOSELY and INTENTIONALLY saying something that is NOT true.

What happens when you confront them with the 'fully bird fossils' found in lower strata? Do they stop pushing their discredited claim? No. For the most part they begin calling the Creationist names and simply dismiss the evidence without giving any reason for doing so.

In searching the internet for Evolution friendly sites that admit birds have been found in layers where dinosaurs are found, you will be looking for a long time.

Remember that I said if ANY ONE of the Theses in this book were correct, Evolution would be proved false. This is the reason why I doubt that you will ever find an Evolutionist or evolution friendly site admit bird fossils have been found with and below dinosaur fossils.

The doctrine of birds evolving from dinosaurs is so strong that so-called 'scientists' will ADMIT this fact when confronted in a one-on-one interview, but will NOT work to correct the displays in museums that tell the unsuspecting public this lie.

Is this REALLY the kind of people YOU want to be following?

There is a 1972 movie called The Poseidon Adventure. The great cruise ship had capsized and a few survivors are making a perilous attempt to get to the engine room so rescuers can cut through the hull and save them.

They RIGHTLY conclude from the obvious tilt of the ship that ONLY the stern of the ship might be above water - if any part is. As such, their only hope was to get to the stern and bang on the hull and hope someone would rescue them.

About half way in their journey - while they had been believing themselves to be the only survivors - they encounter a much larger group of people following one of the ship's officers towards the FRONT of the ship!

The leader of the smaller group fights to get to this officer and argues with him to consider the tilt of the ship - that there is NO WAY for the front to be the direction where salvation will be found! The officer is seriously injured and will not listen, blindly forging ahead - and ALL those in his group follow; scolding the small group leader for delaying their progress.

The small group leader calls out to them as they disappear into the passage way, "You'r going the WRONG WAY!!" - but to no avail. Suddenly, a cascade of water finds its way to the level he is on and fills the passageway where the larger group went.

The leader of the small group barely manages to escape back to his folks that he sent on ahead. He knows that every last one of the larger group is being drowned even as he closes the hatch behind him to slow the water's advance towards his small group.

"Wide is the path that leads to destruction and MANY go that way….." SomeOne infinitely greater than me has said.

Friend, are YOU sure of the one YOU are following?

WHY are you so sure? Because he has a PhD after his name or is wealthy or powerful?

Has someone told you about the ONLY sure way to be saved?

If so - it is time to turn around, don't you think? It DOES make a difference WHO you are following…. an ETERNAL difference.

Theses 86: The Evolution of the modern day horse from a 'dog-like creature from 50 million years ago' NEVER HAPPENED.

The summary below is of a statement NOT from a Creationist - though it would be True even if it had been. No, this statement came from one of the top evolutionists in the world in that day - and he put it in writing in 1953:

"The uniform continuous transformation of the hyracotherium (Eohippus) into Equus, so dear to the hearts of generations of textbook writers, NEVER HAPPENED IN NATURE [emphasis mine}."

George Gaylord Simpson

Life of the Past - page 119

Yale University Press, New Haven, CT

Yet this PROVED WRONG concept is STILL taught today in YOUR public school to YOUR KIDS! I know, I substituted for a science teacher one time and was supposed to tell the kids this lie was actual fact. I was even given an entire kit with the lie all set up so it would be really convincing.

I told the kids the Truth and showed them how they were being lied to about Evolution. Someone must have told the Principal on me, because by the third period - a member of the SCHOOL BOARD came in to monitor what I was saying. I calmly explained to the kids WITH HIM WATCHING the exact same thing I had told the other two classes.

Surprisingly, they did not kick me out that day - apparently not wanting to make a scene or maybe they had no other substitutes. I got to tell the Truth to two more classes - a total of some 130+ kids - of the 75 million kids attending public school that day in America.

However, I WAS banned from EVER substituting for ANY teacher - science teacher or not - ever again. This IS what we are up against, folks. This IS tyranny - and it is EVERY BIT AS DEADLY as the kind of tyranny perpetrated at gun point.

Why? Because young minds are being conditioned by the tens of millions every 'school' day of every 'school' year for the decade and a half these kids go to 'public school'. Then some go on to 'college' where the job is finished by the 'professors' that I rightly call prophets of the Evolution Religion.

YOU feed your child right into the mouth of this machine designed to DESTROY whatever Christian faith you have tried to give them.

They will eventually become the leaders of this and every other country on Earth. They will vote for Communistic and Socialistic policies that reflect that man is nothing but a collection of chance reactions that has little if any value.

It is up to YOU to find out the Truth and then TEACH the Truth to your kids AND defend them when they take on a teacher and get sent to the Principal's office.

It will very likely take homeschooling your kids.

Theses 87: Standard material in any public school 'science' book is the famous 'Peppered Moth' so-called *experiments*.

Two color phase varieties of the same common moth were found in England - and, long story short, as coal was burned with its ash, etc. - the population of dark colored moths is claimed to have risen significantly near places where the tree trunks had this burned coal residue on them and the light colored moth population fell.

The story line goes that birds would see the light colored moths more easily in these coal stained areas and eat them. The 'more fit' black moths would survive and breed - taking over as the majority color phase of this moth.

This is CLAIMED TO BE a very fine example of Evolution.

However, it is NOT Evolution with a capital E - it is actually ADAPTATION and Natural Selection of a critter that ALREADY APPEARED in two color phases!

Even if there were not huge problems with the researcher's methods - which there were huge problems as shown below - these small changes within an already fully developed kind of critter would NEVER give rise to a fundamentally different type of critter. The MOTH would NEVER develop into a BIRD by this mechanism.

Remember also that this moth supposedly arose from the same super amoebae as you did according to the Religion of Evolution.

Some Creationists crow that the experiments were FAKED - and maybe they were. The other side denies outright fakery. Certainly the famous pictures for the books appear to have been faked - for the researcher GLUED the different color moths to the tree trunk.

Still, let us look at an Evolution defense website for their 'explanation'.

Here it is from MillerandLevine.com (a science book publisher that included this material without question for decades, now having to defend itself for doing so).

I have not edited this excerpt and use it under the Fair Use guidelines to make a larger point:

"However, Majerus also discovered that many of Kettlewell's experiments didn't really test the elements of the story as well as they should have. For example, in testing how likely light and dark moths were to be eaten, he placed moths on the sides of tree trunks, a place where they rarely perch in nature. He also records how well camouflaged the moths seemed to be by visual inspection. This might have seemed like a good idea at the time, but since his work it has become clear that birds see ultraviolet much better than we do, and therefore what seems well-camouflaged to the human eye may not be to a bird. In addition, neither Kettlewell nor those who checked his work were able to compensate for the degree to which migration of moths from surrounding areas might have affected the actual numbers of light and dark moths he counted in various regions of the countryside."

Has the Evolution community EVER set out to correct the record in the volume that they trumpeted the 'discovery'? No.

Have they ever sent letters of correction to all of the hundreds of millions of students who were told this data was dependable and believed their information accurate? No.

Has this material been pulled from EVERY 'science' textbook or at least corrections been sent out or directions to not use this part of the book anymore? No.

Do all Evolutionist websites tell their readers the issues with this data? No.

The answer to all these questions is a resounding, 'No'!

In fact, the material is STILL USED today and is STILL PRESENTED as True to YOUR kids. Three of the first four websites that I looked up that were Evolution friendly made no mention of any problem with the research or the conclusions drawn from it.

YOUR tax dollars PAY for these lies to continue to YOUR CHILD!

Theses 88: Of Monkeys and Typewriters a.k.a. mindless typing can produce a work of literature.... really? Does this even need explaining? Apparently so.

We all have heard someone confidently claim that if an infinite number of monkeys hammered away on typewriters for a long enough time - that eventually ONE would type SOMETHING intelligent - such as a sonnet of Shakespeare.

This 'analogy' is called the Infinite Monkey 'Theorem' - and the word theorem should be used very loosely as this story is better characterized as Irish Pub drunken lore - but that very well may insult Irish Pub drunken lore!

Hey, I am Irish - so I can say this; having BEEN to some Irish Pubs! They love to have a great time - but one should hardly bet the farm on ANYTHING that might be claimed in one after about 7 pm. Still, I would RATHER depend on the Irish Pub material than what Evolutionists soberly put out as 'settled scientific fact'.

The Infinite Monkey Theorem was apparently first proposed in the late 1800s - and should have been laughed out of existence right then - but has been seriously put forth in modern literature as late as 2008!

Wow.... Where to begin?

First, ANY TEACHER putting this forth as serious 'scientific' information should have their teaching license revoked FOREVER and sent to bar tend in Ireland.

Why?

Before we can even get to the base assumption of the 'theorem' - riddle me this, Batman:

1) Where did the monkeys come from?

2) How did the typewriters come to be?

3) Who will MAKE and LOAD and ALIGN all the paper and replace the ink

4) Who will feed, care for and replace the monkeys that die so the monkeys are kept at their endless task that long?

5) Who will repair or replace the typewriters that break down?

6) What intelligent being will READ all the nonsense they type to determine that something intelligent has been typed?

Really, folks? This is - or was for decades - actually put forth as serious reasoning in Evolutionist circles!! Yes, sadly - it was - and STILL IS…. it just takes a different form.

MY OWN DAUGHTER heard from a video presentation sanctioned by her public school 'science teacher' that if a car breaks down on the road; it could be fixed by a man THROWING a wrench at the engine numerous times!!!

Again the insurmountable questions arise:

1) Where did the man, wrench and the car come from?

2) Who carries innumerable wrenches in their car?

3) How did the engine HOOD get opened?

4) Are all the wrenches the right size?

5) Is what is wrong able to be fixed by a wrench?

Once a 'teacher' is infected with 'reasoning' such as this and comes to think it worthy of serious discussion: THAT teacher is ready to multiply the FALSE idea many hundreds of times.

Let us take a SINGLE 'science' teacher that has five classes a day each semester with 30 kids in each class. That is 300 kids in a single year that - not knowing better - will believe their teacher. That teacher has a career of 30 years - that is 9000 kids. Each of those kids has five friends that they discuss this with over the course of their lives - that is 45,000 people that a SINGLE TEACHER could influence.

If we do not TAKE BACK our schools; insisting that actual science is taught in science class - our kids BRAINS are doomed to be ruled by proved wrong information and then WE will be ruled by whatever nonsensical policies these eventual leaders of countries concoct and deem to be 'true'.

Remember the old saying, "Be nice to your kids… they will pick your nursing home!" If resources get scarce, these kids may conclude YOU are using too many resources and returning too little to 'society' and establish a policy that you must die or be killed at 65 years old.

Sound crazy? Google the website **rightswriter.com** and ask about former Colorado Governor Lamm...

"DENVER, October 31, 2014 (TheRightsWriter.com) — The former governor of Colorado, who has expressed support for population control and said that the elderly have a "duty to die," has come out against a state amendment that would recognize the rights of unborn children, calling the pro-life measure "a monster."

Never mind that this former governor's OWN MOTHER was already over the age HE determined OTHERS should die. Let that sink in. This man(iac) was in support of killing his own mother in the name of 'population control'. If that is not bad enough - HE was only a few years away from reaching the age where HE would have the duty to die!

Of course, exceptions to this duty to die could be made for those who were contributing greatly to society - like HIMSELF - oh, and HIS mom, since she brought such an enlightened one as he deemed himself to be into the world...

Do you now see how a Hitler could come to think it OK to kill millions of the 'inferior races' that HE had identified as inferior? He was only finishing the building that rested on the foundation of the Evolutionary work of Ernst Haeckel who built that foundation from the blueprints of K. E. von Baer (next Theses).

Theses 89: Humans are HUMAN from CONCEPTION.

This FACT is now known to be fully TRUE. We know this for several reasons; first - aborted baby humans are currently having their organs harvested for research into HUMAN ailments.

These cold blooded customers of murderers like Planned Parenthood PAY big money for HUMAN body parts - not pig or fish body parts.

In the late 1800s, right after Darwin came out with his book, there was a drive to provide evidence backing him up. One of his most ardent supporters was Ernst Haeckel. This man took an idea from predecessor, K. E. von Baer and applied it to human development in the womb.

Did we not just discuss that a single teacher could infect thousands of students that will build on what they are taught?

Like all good Evolutionists - he came up with an incomprehensible but scientific sounding title for his 'theory':

"Ontogeny Recapitulates Phylogeny"

Must be right! It is SOOOOO scientific sounding!

In English this means:

Ontogeny = the complete developmental history of the individual organism

Recapitulates = re-enacts during embryonic development

Phylogeny = the complete evolutionary history of a group of organisms

Got that? They hope you did not either.

Evolutionists MUST baffle with B.S. when their ideas cannot stand up to scrutiny.

I cannot tell you how many times in my online arguments with Evolutionists that - when they cannot defeat my logic - they simply say, 'You just do not understand how Evolution works!' - to which I reply, 'Apparently YOU do NOT understand how it works, EITHER! - for if you did; you SHOULD be able to convincingly show me WHY what I have stated is wrong - and you SHOULD be able to put your concepts into simple enough terms that intelligent people would be able to grasp.'

Here are some excerpts representative of what can be found by a simple search regarding Haeckel's work:

"The idea that embryos of different organisms look similar is not foreign…. knows that chicken embryos are almost identical to the embryos of humans."

OK, stop! NOT TRUE!!

Haeckel's drawings were completely MADE UP as the website **evolution.berkeley.edu** states, "Haeckel was so convinced of his Biogenetic Law that he was willing to bend evidence {read that: LIE} to support it."

We could go on for a long time - debunking literally each line of the excerpt - but my goal is to give you the determination to learn how to do it for yourself.

Just one more and we shall move on to the next Theses:

"In light of the newly found process of evolution, it was determined that mammals descended from reptiles which descended from fish. It was hard to ignore the observed similarities in the embryos of these groups. All three developed gill slits..."

STOP!

Human embryos have now been shown to NEVER develop gill slits! Yet this LIE is used throughout the abortion industry to reassure the young girl that she is only ending the life of a fish at this stage of fetal growth - not killing a baby human.

Indeed, it is now known that the baby's heart starts to beat just 18 days after conception - often BEFORE a woman even knows for sure she is pregnant.

Ernst Haeckel proved WRONG:

"However, almost immediately after the famous words were uttered, 'ontogeny recapitulates phylogeny' in the strictest sense was proven false. It is now firmly established that ontogeny does *not* repeat phylogeny. Ontogeny repeats ontogeny, with variations," (Pittendrigh 352-53). It is now widely accepted that Haeckel's statement was false..."

Yes, it is KNOWN in Evolutionary circles that Haeckel was dead wrong (or as Creationists have shown - was lying the whole time).

Haeckel's 'work' was a major foundational cornerstone to the early eugenics projects and ultimately found its way into the heart and soul of one Adolf Hitler…. on which foundation he built his ranking of humans…. that he used to justify extermination of millions….

Do you STILL want to stand idly by and let the brainwashing of millions across the world continue? How long will it be before ANOTHER Adolf Hitler gains control of a powerful enough government to plunge the world into world war?

Yet this grave error has been allowed to continue to be relied on by the general public. Worse, those few in the Evolutionary camp that still embrace this 'fake science' are interviewed and published as authorities on the subject and printed in the 'fake news' outlets! I saw in Life Magazine somewhere around 2008 a front page showing Haeckel's work and article claiming it was true….

Summary of these last few Theses:

By now it is plain that Evolutionists are not just wrong about the evidence. They are not proposing valid and possible alternative understandings about how life came about - and in many cases know they are not.

It should be crystal clear by now that Evolution is a Religion and NOT science - making it the prophesied '…science, falsely so-called..' that Paul predicted would come in the end days when God would send a powerful delusion on the un-Godly.

Theses 85 through 89 detail KNOWN LIES that have been exposed for DECADES that are STILL being taught or promoted as true to YOUR CHILDREN by public schools and universities around the world.

YOU are not just paying taxes to promote these lies, but ALSO are paying sky-high TUITION to have your kids brainwashed. Then you wonder why the Christian son or daughter you sent to college comes back as a Socialist, Communist or Atheist (if not all three)!

The ONLY way to stop this is to STOP OFFERING UP your children on the ALTARS of the FALSE GOD of the Evolution Religion - called secular colleges and public schools.

Theses Ninety through Ninety Two: Geology Shows There WAS a World Wide Flood that Covered the Entire Earth: Identifying the Source of the Water, Grand Canyon Discussed, and the Mars Flood

Premise: The entire Earth shows evidence of layers of sedimentary rock many hundreds or even thousands of feet thick; therefore: Since sedimentary rock REQUIRES water for deposition - it is logical to conclude there was a world wide Flood in Earth's past.

Theses 90 - 92: These three Theses will be handled as a group - since they are interconnected. Let us go to Mars first - then return to Earth.

In 2013, in an effort to justify the billions of dollars spent by NASA on various space exploration projects - SOMETHING dramatic HAD to be found each trip to justify still more spending - and so something was always found.

Here is one headline from Time Magazine (that bastion of Truth - now outed as part of the fake news cabal):

The Great (and Recent) Martian Flood
The Red Planet may be home to a vast ocean of water—and it's straining to break free
By Jeffrey Kluger March 11, 2013

Given the headline above and breathless story that followed: WHY may I ask is it SO impossible for there to have been a planet engulfing flood here on Earth - where water still covers 70% of our planet?

A quick survey of several websites that come up when you search for the situation here on Earth under; 'Water/Land Ratio If The Polar Ice Caps Melt Completely' - reveals a general consensus that about 80% to 85% of the Earth would be covered with water if the ice caps melted completely.

These 'scientists' apparently have NO TROUBLE believing in planet inundating events on Mars - where NO STANDING WATER WHATSOEVER has actually yet been found. We do see the Martian 'polar ice caps' - apparently consisting of water-ice as well as dry-ice (frozen carbon dioxide) - but have NOT found any physical lakes or oceans on Mars.

Let me get this straight:

On Mars, where there is no known free standing water - this article claimed a planet sized flood most certainly happened - AND RECENTLY! Why? Because we see SEDIMENTARY ROCK formations and other ' must have been shaped by water' features are evident.

YET on Earth; where we have LOTS of free, non-ice bound water with even more water bound up in ice at the poles - AND literally everywhere we see SEDIMENTARY rock layers AND many water carved features: Evolutionists say there is not even the remotest possibility that a world wide Flood occurred!

It would seem that it is because this world engulfing flood on Earth has a NAME attached to it that Evolutionists MUST say it never occurred.

What is the name of that Flood? Hmmm….. it was on the tip of my tongue….. Oh, yes!!

Noah's Flood.

This Flood is TIED directly to Creator God's WRITTEN WORD and is associated with JUDGMENT on the unGodly. The Flood showed that God is capable of meting out Judgment and WILL bring that Judgment after His amazing patience is exhausted.

It also shows that the Bible is TRUE from beginning to end - for ONLY the Bible tells the story of HOW such wickedness came to be that would necessitate God judging in such a dramatic way.

Of course, in our day - wickedness not seen on the Earth SINCE the days just before the Flood are everywhere. The phrase '…EVERY thought of their minds was ONLY evil, continuously…' is the most appropriate descriptor of much of modern mankind's thoughts and actions. Man today not just invents new kinds of evil - but rejoices in the kinds of evil other men invent AND also takes great pleasure in persecuting ALL who attempt any restraint however minimal.

One example is enough to demonstrate this:

God created male and female - MAN and WOMAN - Adam and Eve (not Adam and Steve). The 'transgender' movement now claims up to 32 sexual orientations! There are even PEOPLE that dress up like ANIMALS and have others walk them on leashes, etc. There are PEOPLE marrying not just the same sex (abomination enough in God's sight) - but they are marrying the ocean or the trees or the sky!

Moving on....

What about Grand Canyon?

Evolutionists claim it was slowly carved over millions of years by that little river now found at the bottom of the canyon. A careful look at many aspects of the canyon show that it was carved RAPIDLY - the result of a breached natural dam that once held back Grand Lake and others include Hopi Lake. The article at Answers In Genesis (AIG) factors in high rainfall occurring after the Flood as a contributing cause.

I have stood at the starting end of the canyon where a very obvious natural dam breach is clearly visible. Heading into Utah, what seems to clearly be a now dry lake bed has cliffs several hundred feet high on both sides - with the sandy desert dunes in between these cliffs clearly being the former lake bed. The author of the AIG article says the volume of water held in these huge - but now dry - lakes to be at least three times the volume of water now in Lake Michigan.

The elevation of the land at the beginning of the canyon is actually well BELOW other parts of the canyon! If that river did cut the canyon - it would have to have flowed UPHILL to do it!

The only way Evolutionists accomplish this feat of magic is to imagine a simultaneous, very slow uplift of the land AS the river eroded through it. The ONLY REAL WAY anyone can prove water to temporarily flow uphill - WITHOUT a pump making it do so - is for there to have been some other great force behind the water. A natural dam breach would provide just such a mechanism.

The Evolutionists know this fact and how damaging it is to their claim of slow erosion over millions of years; so they simply claim that this section of land 'uplifted' after the fact. They have no proof of this claim for they did not see it happen - but they cannot give any credence to Noah's Flood - so they simply say it happened.

Further, the AIG article points out that the Evolutionists have changed the 'age' of the canyon many times over the decades - now saying it is only 5 million years old instead of 70 million. In all the crime movies I have seen - real or made up: A sure sign a suspect is lying (or a witness is unreliable) is that they constantly change their story…..

The timeline of the Evolutionists - 70 million or 5 million years - is claimed for only ONE reason.

That reason is to cast doubt on the Bible's timeline of Noah's Flood. If they ever admitted that something as massive as Grand Canyon could form quickly - and further admitting they do NOT know the age of the canyon - then many people's minds currently shackled by the information required to be put out by Park Rangers as fact would be broken free.

Once people realize they have been lied to or purposely misled ONCE, they NEVER hold the rest of what that source tells them as fact again. They will FOREVER be suspicious of the other side's 'Rest of the Story'.

This is WHY the Bible MUST be actually TRUE from Genesis to Revelation. It is also why Evolutionists and Atheists so desperately try to find even a single error in it. Even if that error was tiny and insignificant - they would claim the right to disbelieve it all. It would not matter to them that in this book I have shown 95 entire Theses indicating Evolution is a LIE - most would cling to their RELIGION and stuff their fingers in their ears.

Of course, Creationists did not see the canyon form either - and there are some features we do not yet fully understand - BUT an objective look at ALL the evidence currently available clearly aligns much better with the Noah's Flood aftermath scenario.

Grand Canyon's layers apparently were quickly laid down and hardened into rock (possibly soft rock) before being suddenly eroded when the Flood waters began to run off and then the natural dam burst. We know the layers were laid down quickly because of the nearly pure uniformity of the deposits consistent with sediments deposited by rapidly flowing water.

Finally, let us address one of the Evolutionist's favorite questions: If there was a worldwide Flood: Where is all the water?

Talk about not being able to see the forest for the trees!

ALL that water is STARING THEM RIGHT IN THE FACE!

It IS in the OCEANS now!! - with a large percentage locked up in thick sheets of ice at the poles.

Just for a change, let us take God's Word as an accurate account. Genesis says there was water in a 'Great Deep' apparently somewhere in or under the Earth's crust. There was also water on the surface of the Earth. The also seems to have been water 'above the Earth' - which many claim may have made Earth into a type of terrarium accounting for conditions that led to the great lifespans of people recorded in Genesis - but that is a topic for another book.

An analysis of the Hebrew indicates what we read as the English word 'mountains' in the first part of Genesis were really just 'high hills'.

Now let us examine some FACTS. The average depth of the ocean today is about 12,000 feet. The average elevation of all land currently above sea level is only about 2600 feet.

Now, let us smooth out the Earth's mountains and fill in the valleys to make a perfectly smooth Earth. What would be the result?

The Earth - TODAY - would be completely submerged by more than a mile of water!

What about Mt. Everest? Fossils of marine creatures are found on top of Mt. Everest as stated previously. There HAVE BEEN uplifts of large sections of land - with Mt. Everest being the most dramatic. However, even the land that is now Mt. Everest was once under water for a long enough time for marine critters to grow and flourish.

The Bible tells us where this water came from - under or in the crust of the Earth. The Bible tells us about a Flood that is completely confirmed by any objective look at the evidence.

I choose to believe SomeOne or someone until THEY prove themselves to not be trustworthy by LYING or MISLEADING or if they engage in unwarranted speculation.

God's Word has NEVER been shown to be in error - when the original language is examined and the original intended audience's cultural situation is considered.

On the other hand - I just got done documenting numerous **PROVED WRONG IDEAS** and **OUTRIGHT LIES** of Evolutionists!

The Bible DOES contain some admittedly hard to understand concepts - like angels and miracles - BUT in another of my books (Lumpy Oatmeal with Raisins and Cinnamon!) I give a plausible explanation of these phenomena.

Many CLAIM there are errors in God's Word - and I have challenged many online to give me their best so-called error. To date, NOT ONE of my opponents has shown a TRUE error WHEN the original language is studied AND the original cultural setting is considered.

Let's see:

On one hand I have God; Who has never lied to me or misled me in any way

- versus -

Evolutionists and Atheists who have CONSTANTLY lied to me or withheld critical evidence from me or engaged in wild speculation trying to pass it off as science.

Which should I trust, follow and order my life by?

I do not know about you, but I am going with God.

Theses Ninety Three and Ninety Four: What Evolutionists Say When They Think No One is Listening

Premise: If Evolution is really True, then the purveyors of it SHOULD be confident in their view even when in unguarded situations. If these 'experts' show they have doubts when talking to each other, then we are justified in having at least the same doubts.

The quotes used in these final few Theses are taken from The Revised Quote Book published by the Creation Science Foundation under the Fair Use guidelines. I encourage anyone reading this book to purchase this very wonderful resource!

Theses 93: The Creation vs. Evolution debate has gone on in earnest for more than a century in this country; therefore it should come as no surprise that there is no shortage of things said on both sides in all that time.

For the last 65 years or longer, the Evolution side has had the public megaphone in the form of a near monopoly of favorable major media coverage and near total control of the public school 'science' curriculum. Over 90% of American kids are said to attend public 'school'.

This begs the question: Given the advantages for Evolution's view stated above - WHY, when polled, do nearly 40% of all people STILL believe in a literal or a 'mostly literal' Creation scenario? Further, why do up to 85% of people want BOTH views to be taught in public schools?

The reason people do not believe the Evolution Religion is simple! ANY attempt to make nothing-to-molecules-to-man Evolution plausible just does NOT make any sense to people that are still able to think for themselves!

For those who remember the Peanuts cartoons, this classic applies: Charlie Brown and Linus are putting together a puzzle. They look confused or frustrated - so they begin jumping up and down on the puzzle! Lucy has been watching them and raises an eyebrow. Linus tells her, "Its a puzzle… if the pieces don't fit…. we MAKE THEM FIT!!"

People who doubt the Evolution story instinctively understand that the information they have been forced to regurgitate to pass exams just does not fit! They may not be able to tell you exactly why they reject what they had force fed to them in their local indoctrination center; but they still reject it or at least think it reasonable to allow a Creator to be considered as a possibility.

A super majority of people - in some polls up to 85% - WANT either Young Earth Creationism taught exclusively or BOTH views taught in public schools!

Totalitarian school boards however thwart every attempt to teach Creation and if they are not able to smack it down themselves; the ACLU (the 'Atheists, Communists, Liberals Union') threatens to sue, claiming that teaching Creation = teaching a religious view.

The fact is that religion is ALREADY being taught in science class at tax payer expense - the Religion of Evolution!

The next two Theses will document some verified statements by Evolutionists obtained from venues where they thought no one was listening or no one would ever see what they had written - except a fellow traveller.

In these quotes you will see WHY very many people just do not buy Evolution even after having it drilled into them throughout their school years.

Quote Set 1: Is Evolution a form of Science or is it Religion?

"…evolution became in a sense a scientific religion, almost all scientists have accepted it and many are prepared to 'bend' their observations to fit in with it." H. S. Lipson, Professor of Physics, University of Manchester, UK.

"With the failure of these many efforts science was left in the somewhat embarrassing position of having to postulate theories of living origins which it could not demonstrate. After having chided the theologian for his reliance on myth and miracle, science found itself in the unenviable position of having to create a MYTHOLOGY of its own… (capitalization mine)." Loren Eiseley, Ph.D Anthropology from 'The Secret of Life' in The Immense Journey 1957, page 199.

"....{my book} will be grievously too hypothetical.... alas, how frequent, how almost universal it is in an author to persuade himself of the truth of his own DOGMAS (capitalization mine)." Charles Darwin in a letter to a colleague found in John Lofton's Journal, Washington Times, Feb. 8, 1984.

Even Charles Darwin - Evolution's founding father - ADMITS to a friend in a private letter that his work is but DOGMA. Anyone familiar with religious terms knows that a dogma is a RELIGIOUS term for a strictly held article of FAITH!!

Though these quotes are from more than 50 years ago - the problems they opine about have only grown worse with the greater information we have gained in those five decades.

Quote set 2: What actual facts do Evolutionists have on their side?

This question is answered by Dr. Colin Patterson, Senior Paleontologist at the British Museum of Natural History in London during his keynote address at the American Museum of Natural History, New York City, Nov. 5, 1981:

"One morning I woke up and something had happened in the night, and it struck me that I had been working on this stuff for twenty years and there was NOT ONE THING I KNEW about it.... Either there was something wrong with me or there was something wrong with evolutionary theory…. for the last few weeks I've tried putting a simple question to various people…. 'Can you tell me ANYTHING you KNOW about evolution, any ONE THING that is TRUE? I tried that question on the geology staff at the Field Museum of Natural History and the only answer I got was SILENCE. I tried it on the members of the Evolutionary Morphology Seminar in the University of Chicago, a very prestigious body of evolutionists… all I got was SILENCE for a long time and eventually one person said, 'I do know one thing - it OUGHT NOT TO BE TAUGHT IN HIGH SCHOOL' {all capitalization mine}."

This was one of the world's leading Evolutionists speaking to a very important gathering of his peers! Why then - over 35 year later is it STILL being taught to millions of kids as if it were factual, real, settled science?

By the way, I visited the very famous Chicago Field Museum of Natural History a few years ago. I was wearing (on purpose) my best Creationist T-shirt that refuted Evolution on the back.

We had only been there about a half hour when we noticed a man in a civilian suit seemed to be following us! Every exhibit we went to he tagged along, trying to be unnoticed but failing badly. It was the 150th anniversary of Darwin's book - and they had an entire display dedicated to him. Since there was an extra charge, I went in but my family went to lunch. He came in on my heels - and I noticed that he did not have to pay… hmmmm.

As I looked at each display, I shook my head in disagreement and even struck up conversations with strangers pointing out the problems with what was being put out. The man would draw near each time I said something.

After finishing the walkthrough, I joined my family for lunch. The man took the opportunity to strike up a conversation with me! He did not introduce himself as a museum employee or anything, but did say he had been reading my T-shirt and listening to my conversations with others - and wondered why I would even come to the museum if I disbelieved Evolution?

I asked him if I was in any sort of trouble for peaceably exercising my right to free speech and he said no as long I as did not harass anyone; giving away that he had been monitoring me! In between bites I took him to school regarding the fallacies being portrayed as fact to the public. In each case, he could not indicate that my assertions were wrong.

He did not have us kicked out - since we were not causing any kind of a scene - but after he left, I noticed a security guard shadowing us until we left for the day. Apparently Evolution is on such shaky ground that even a Creationist T-shirt is a threat!

Theses 94: The other main reason so many people remain unconvinced of the claims of Evolutionists is that it simply fails the 'sniff test' of Truth. ANY even cursory examination of Evolution tells the examiner that something is VERY ROTTEN in Denmark!

All capitalization and emphasis in the following quotes is mine. Those who push Evolution have said the following:

Quote set 3: Are there ANY facts that PROVE Evolution happened?

"Scientists who go about teaching that evolution is a fact of life are GREAT CON-MEN, and the story they are telling may be the GREATEST HOAX EVER. In explaining evolution, we do not have ONE IOTA of fact." Dr. T. N. Tahmisian, Atomic Energy Commission USA in the Fresno Bee, Aug. 20, 1959 as quoted by N. J. Mitchell, Evolution and the Emperor's New Clothes, Roydon Publications, UK 1983.

"...evolutionary trees that adorn our textbooks have data only at the tips and nodes of their branches, the REST IS INFERENCE.... NOT the evidence of fossils." Stephen Jay Gould, Professor of Geology and Paleontology at Harvard University in Evolution's Erratic Pace in Natural History, volume LXXXVI(5); May 1977, page 14.

"Evolution requires intermediate forms between species and paleontology does NOT provide them." David B. Kitts, Ph.D.

Even though ALL these misgivings about Evolution are kept from students forced to learn Evolution as if it were factual - the gaping holes and obvious contradictions make selling this soup sandwich nearly impossible.

What the public schools DO succeed in is forcing kids to regurgitate Evolutionary doctrine long enough to get their grade - because not very many students will risk failing by arguing.

This is why I was SO PROUD of my daughter when she wrote this while she was taking a test on Evolution in high school:

"I do NOT believe in Evolution. If you have a problem with that, call my dad. His phone number is….." Later in the same exam, she answered another question, "See question #2" and left it blank!

Her teacher NEVER called me and GAVE HER CREDIT for both unanswered questions! I had a reputation at the school board by this time and he apparently did not want to tangle with 'dad'.

My daughter's resistance was every bit as brave as that of Rosa Parks when she refused to give up her seat and move to the back of the bus! This is NOT an overstatement just because she is my daughter. The SLAVERY that Rosa's forebears suffered and the blatant discrimination Rosa rebelled against was given 'scientific cover' by Charles Darwin's book by allowing 'superior' people an excuse to enslave 'inferior' people groups.

Though slavery had long been abolished in America by the time Rosa bravely stood her ground, the OPPRESSION she was resisting sprang from the same vile root that spawned slavery - Evolution.

Need proof?

Take a look at EVERY 'artist depiction' of 'primitive man'. What is the color of the skin selected by the artist for these 'brutes'? What kind of hair do they have? Contrast the artistic rendering of 'modern men' such as Cro-Magnon man. He is ALWAYS depicted as having lighter skin and hair more like a Caucasian person. What subtle message is being sent here? - that as man progresses, he somehow morphs into light color skin and non-Afro hair….

Compare what the Bible says of ALL people… God's Word tells us we are ALL "…born of ONE blood…" - and ALL equal since "…in Christ there is no slave nor free, male nor female…" Racism and Sexism ARE logical off-shoots of the Evolution Religion.

Slavery affected a large group of people. Evolutionary indoctrination affects nearly ALL people - including those that have courageously thrown off the chains of slavery.

While my daughter's act of courage will likely never be known outside the pages of this book; it will take many millions of 'Rosa Parks' type actions to destroy the stranglehold Evolution has on the United States and most other nations of the world.

Where are all the courageous men of God willing to face some mild criticism when they challenge those who claim their God is a liar?

The martyrs of old went to their DEATHS - often times joyfully - to defend what was at that time just a verbal account! The Bible had not even been fully written or put together into a single Book at that time!

As a man speaking to men for one moment: YOU are given the SPIRITUAL responsibility for YOUR family. YOU will have to give an account for YOUR action or inaction in this arena.

In another of my books I wrote a poem honoring the martyrs of King Jesus. Here it is and the poem may be shared freely as I have released my copyrights to it.

Confessor Today?

As I stir my designer coffee,
Settling into my reclining chair
Lamp on so I can better see
Remotely adjusting conditioning of the air

Opening the ancient martyr book
Purchased from an online store
I transport back in time for a look
At the Faith of the true Christian confessor

Blandia! - a slave: Why did she suffer so?
'The Name of Christ' I read
For in Rome two thousand years ago
A capital crime was it merely to believe

The capricious 'gods' of Empire Rome
Could bear to have no rival
One Who made master and slave, men and women
All before Christ: equal!

'They must be killed! - and horribly so!'
An example to one and all
"Swear Caesar is God - or off you go
To the arena where you will fall!"

With dozens more, Blandia heard her fate
"Fed alive to beasts or burned!"
Dragged through that gate
"Deny Christ and live!" - a last option did she learn

Those watching her torture greatly marveled
As her torment went on three days
'Til their blood lust a short Roman sword filled
Even still - no denial did Blandia raise!

It is said of even that hardened crowd
"No woman ever suffered so!"
Then breaking trance, I reached out
To find my latte had grown stone cold…

Accounts of thousands more rolled on
Through centuries up 'til our own day
Tyrants freely shedding righteous seed-corn blood
For God alone they would obey!

Lord Jesus! Please forgive my sin!
When your witness I am shy to be
Tomorrow before my skeptic 'friends'
Put Blandia's faith deep in me!

If you need help answering questions, here is my email: **kingjames2bible@gmail.com**

I will answer as many questions as possible - but there is literally limitless information on the internet by people far smarter and much more qualified than me.

PLEASE look there first, as I am but one man in this fight. I will include a list of resources that I have used over the decades. Most are still available or can be found with a little searching.

Theses Ninety Five:
Evolutionists DO NOT Really Believe Their Own Concocted Story and Betray Their Unbelief By Their Own Words!

Premise: Similar to Theses 93 and 94; IF we find Evolutionists openly admitting that they know or suspect major portions of their 'theory' are impossible unless one willingly bursts through the limits of credulity; THEN we may confidently REJECT EVERYTHING the Evolutionists say.

I will say that again:

You are fully justified in REJECTING absolutely EVERYTHING any and all Evolutionists put forward.

My online critics howl loudly that I think I am right and all the 'scientists' are wrong. I correct them and say that it is NOT me that is right - but God's Word and the One Who wrote it that is Right. I am but an all to often imperfect messenger.

Further, not all the scientists are wrong - but EVERY scientist that has sold his soul to the Religion of Evolution IS wrong. Every last one of them! It does not matter at all that a majority of so-called scientists have surrendered their objectivity and intellect to Evolution. Majority opinion and a $1.95 will get you a basic cup of coffee. In other words, majority opinion is of NO VALUE at all.

Since Creationism as a Religion is the exact OPPOSITE of Evolution Religion AND the Evolutionists themselves do not find molecules to man Evolution credible AND there is no third alternative:

We CAN BE POSITIVELY SURE:
+ Creator God exists
+ He did Create the entire Universe and us
+ His Word is surely True and without error
+ We are sinners needing His Remedy for our malady
+ Eternal Life is possible
  + Is the Gift of a Just, Holy and Merciful God
  + Who WANTS us to be Saved by His Only Son's Finished Work
    + of Redemption on the CROSS!

So just exactly how confident are Evolutionists in their Theory? Let us find out.

Theses 95: The very best Theses is saved for last!

Evolutionists passionately engage and argue with Creationists in nearly every imaginable venue.

Given human nature, this is understandable. Men - though wrong and even knowing they are wrong - will very often PUBLICLY fight for their error rather than admit they are and have been wrong.

Ladies, there are some of you out there on the Evolution side, too - and just as stubborn as any guy I have tangled with!

I am sure there is some psychological term for this - insanity, psychosis, schizophrenia - whatever; but WHEN they are NOT engaging the 'enemy' (though we are really their best friends concerned for their eternal souls more than they are); they say things that PROVE they really do not believe what they put forth for others.

Do not get me wrong: They DO HOPE it is True.

They DO HOPE there is no such thing as 'sin' and no God and no Judgment for their sins. Many if not most Evolutionists that have been 'in it' for a while really do understand that the possibility of their being right is so remote that only a desperate fool would put their faith in it. Yet they publicly still defend 'the Theory'.

For some, it is because they fear - and rightly so - being ostracized from their prestigious group of fellow travelers. If they revolt, they WILL be shunned, fired, smeared and worse. For others, they simply cannot admit that they have been so very wrong - especially when they would have to admit the Creationists were right. Still others do not want to give up their sinful lifestyle and / or face the fact that one day they will die.

There is even a new group that believes they will soon be able to 'upload' their brains into machines and 'evolve' into the next being. They are simply following their leader in his vain attempt to 'change over time' into something greater than God.

The following is a valid quote from a very famous Evolutionist outlining the odds of nothing-to-molecules-to-amoebas-to-worms-to-lemurs-to-monkeys-to-man being actually true.

Read the full statement a few times and let what the man says sink in deep. If he - a LEADING FIGURE in the Evolution community - can say this with a straight face - all I can say is 'Wow!' I never knew delusion could be so strong.

The quote is again taken from the booklet, 'The Revised Quote Book - Quotable Quotes on Creation / Evolution by Leading Authorities' published by the Creation Science Foundation.

The booklet is nothing more than the validated words SPOKEN and WRITTEN by Evolutionists with very little and often no commentary by the Creationist editor.

Here is the quote of all quotes where the Evolutionist equates the odds of higher life forms being the result of Evolution as comparable to:

"...a tornado sweeping through a junk-yard might assemble a Boeing 747 from the material therein..."

Sir (as in knighted by the Queen of England) Fred Hoyle
Professor of Astronomy at Cambridge University
quote from article titled 'Hoyle on Evolution'
Nature volume 294; Nov. 12, 1981 page 105

The magazine Nature, like National Geographic, is completely sold out to the Evolution Religion.

It is a very great honor for a person to be tapped as an 'expert source' for one of their articles. This is 'Evolutionists talking to each other' - never suspecting that someone would question their basic premise.

THINK about what he has just said!

A tornado - a source of chaotic energy that ALWAYS in past experience has represented DESTRUCTION of already intelligently assembled things - might someday ASSEMBLE something as complex as a modern aircraft!

A tornado generally MOVES rapidly across an area - maybe lasting a FEW MINUTES in any given small area such as a junkyard.

This man says the chances of Evolution being responsible for the existence of man are AT BEST only this good!

However, let's look at what this man is really saying.

Evolution supposedly took hundreds of millions of years to get to man AFTER the first super-amoeba magically appeared. This imaginary 'creative tornado' would have only minutes to do it's work of assembling a seriously complex airplane - and that is if ALL the parts were present in said junkyard - AND in working order!

If you compare the timelines, those few minutes of a tornado's presence would essentially be instantaneous creation compared to Evolution's billions of years of mutation adding to mutation timeline!

This man betrays that he actually BELIEVES that 'life by instantaneous creation' is not only possible, but is the ONLY way he can figure out how life 'arose'.

The only thing he cannot bring himself to do is give the credit to Creator God - so he credits a tornado instead!

What is more, Nature magazine put this out as their 'gospel' - to their congregation of tens of thousands of believers…. ummm, I mean subscribers.

There seems to have been no outcry to fire this man for saying this - therefore by their silence and lack of a better explanation; they all agree!

Let us look at two more quotes regarding this same topic of the origin of life without Creator God as a possibility.

Dr. Leslie Orgel; biochemist at the Salk Institute in the article, 'Darwinism at the Very Beginning of Life' in New Scientist, Apr. 15, 1982 on page 151:

"The origin of the genetic code is the most baffling aspect of the problem of the origins of life and a major conceptual or experimental breakthrough may be needed before we can make substantial progress."

Yet another quote from Richard E. Dickerson, Ph.D at California Institute of Technology in an article in Scientific American, volume 239(3), Sep 1978 on pages 77 and 78:

"We can only imagine what probably existed, and our imagination so far has not been very helpful."

Both of these 'scientists' freely admit that they DO NOT KNOW in the slightest degree how life 'arose' without Creator God - and yet both publicly STILL consider Evolution to be 'settled science'!

This is why Creation IS no more of a religious construct than is Evolution.

Whichever view you choose can ONLY be believed - as BOTH views are RELIGIOUS in nature. Therefore EVERY human being is 100% religious regarding the topic of Creation versus Evolution.

The Inescapable Conclusion:
There is a Creator God
He IS the God of the Bible

I have outlined only 95 of many hundreds if not thousands of Theses that could have been put forth that utterly show that Evolution is a completely wrong accounting of all that we know to exist.

Since Evolution is wrong and there is only one other plausible proposal on the table - and that proposal is exactly the opposite of Evolution in every respect: Creation - recent Creation some 6000 years ago - is the correct view:

We have examined evidences from:

+ The very limits of the known Universe
  + Galaxy rotation and still very structured appearance
  + Short Term Comets
  + Star deaths with the absence of plausible star births
+ Our own Solar System
  + Our own Sun and Moon
  + Some of the other planets and their moons
+ Our Earth's geology
  + River Delta size
  + Grand Canyon formation scenarios from both sides
  + Erosion rates of mountains and supposed ancient fossils in them

- The general failure of Radiometric Dating
  - Same critter yielding two wildly differing dates
  - C-14 still in supposed hundred million year old coal
- Chemistry
- Biology
- Logic and common sense
- OOPARTS
- Natural phenomena that Evolution cannot explain
- Where all the water is that once covered the Earth
  - Marine life fossils atop all mountains including Everest
- The MANY frauds and misinterpretations on Evolution's side
  - The ENTIRE missing chain from monkey to man
- Evolutionists admitting their disbelief in 'the Theory'
- and so much more

Some in the Creationist camp will wonder why I have not covered their favorite Theses. Some that I did not include though they are certainly as valid as any Theses found herein are:

- Polystrate fossils - fossil trees that span 'million year' layers
- Rapid formation of stalagmites, stalactites and flowstone
- Rock layer angles indicating they formed under water
- Soft tissue preservation in fossils proving rapid fossilization
- So many chemical reactions in cells that require design
- Evolutionists charge of poorly designed human eye shown wrong
- Debunking the Atheist's claim that all animals could not fit on Ark

+ Frozen baby mammoths dying of dust asphyxiation not drowning
+ Dozens of 'living fossils' showing NO change over supposed eons
+ Etc., etc., etc., etc., etc……… etc!!

Literally EVERY SINGLE known fact or find - when properly understood will be best understood from a Creation perspective.

In Charles Darwin's day; PERHAPS - lacking all that humans have discovered and proved since then - would be some excuse to entertain the idea that nothing could produce something and a super amoebae could morph into a man - but not today. Today there is NO excuse whatsoever to hold any other view except Young Earth Creation.

In Darwin's day as Russell Grigg documents in his article Darwin's Mystery Illness for Creation Ministries International Sep 1995; he convincingly argues the Father of Evolution suffered from some type of anxiety caused psychosis. He shows that Darwin did not know of:

1) Gregor Mendel's work on genetic trait transfer by precise mathematical ratios, Darwin thought blind chance accounted for heritage

2) James Joule, Lord Kelvin and their peers were just developing the Laws of Thermodynamics, the first of which gives the first two Theses of this book and shuts the door on Evolution by themselves

3) Louis Pasteur was just beginning his work that proved the Law of Spontaneous Generation absolutely wrong and the Law of Biogenesis correct - which eliminated the super amoebae idea

4) The fossil record was barely investigated - so Darwin's own claim that if myriads of transitional fossils were not found, his work would be proved false - by his own criteria, his Theory has been disproved

According to Mr. Grigg; Darwin himself referred to his work as "…my accursed book…" and held off publishing it for nearly two decades! He only pulled the trigger when another appeared to be ready to beat him to the punch.

As stated in the beginning of this book, if ANY SINGLE ONE of these Theses is actually True - or if any valid Theses not covered by this book are True: Then Evolution is FALSE. Since one of the ONLY TWO possible concepts is shown false, by default the other must be True. Therefore let me state the situation most clearly yet again:

Since no-God Evolution is FALSE, then its opposite; Creation by God, must logically be True - since there are no other valid options known to man.

Since every one of the Theses is True, there is absolutely no possibility that Evolution or it's claimed timeline are True.

A majority of mankind OFTEN falls for a lie that is later proved to be a lie. Much of mankind has fallen for a lie once more - and it is a big LIE - a lie so pervasive and destructive it is worthy of the father of lies being its author as has been pointed out in these pages.

This lie currently controls EVERY public school in America and most colleges - and therefore has at least clouded the thinking of the vast majority of otherwise intelligent people.

Even though most still go on to reject Evolution personally, since they do not know the extent of the lie - many have not done what it takes to eradicate it from their lives as much as possible. Those that have are called Creationists. Those that have mixed a degree of Evolution with Creationism are called Theistic Evolutionists - God (or god) did it using Evolution.

Since human action springs primarily from one's level of belief - Creationists lead the charge to promote the Truth. Atheists fight them tooth and nail. Both sides are True Believers in their concept - even though only one can actually be correct (Creation).

Theistic Evolutionists sit in no man's land like the lukewarm water that Jesus says He will spew out of His mouth!

It is time - RIGHT NOW - to firmly decide which side you are on!

Elijah confronted the false religion of Baal in his day. He was outnumbered by the priests of Baal 450 to 1! If he failed in his mission he was a DEAD MAN - for King Ahab and his wife Jezebel had already murdered nearly all of the other True prophets. While the situation is not quite that bad today - academic execution most certainly still happens to those who dare question Evolution in the slightest degree.

Elijah asked the confused and compromised Israelites, "How long will you sit there in indecision?! If Baal is God; serve him. If God is God; serve Him!"

With this book; I - a poor shadow of the great prophet Elijah - issue the same challenge to everyone reading this book:

"If Darwin's Evolution is God, serve yourself (for there is no God). If All-Mighty Creator God is God - serve Him!"

If after reading this book you now know that Evolution is not just impossible, but also a complete LIE - then take a stand! First: Arm yourself and your spouse with Creationist information and do whatever it takes to save your CHILDREN from being indoctrinated into this FALSE RELIGION of Evolution in public school.

Next: Enter the fight! Get more information and pass it out; talk your pastor into hosting a Creation event; run for school board or even start your own Creation ministry.

If you WILL NOT - for this is a matter of will - remember what comes of unrestrained Evolution if it gets married to political power of the State: Hitler's Germany, Mao's China, Lenin's Russia, Planned Parenthood's 'clinics' where more than 100 million pre-born worldwide have been killed.

To the extent we sit on our hands and fret; allowing this lie to fester - from the pool of Evolution compromised folks come our current day leaders. We need these people to engage the Creator Who made us and lead us in **PRAYER** back to Him! Very rarely do our leaders do this publicly anymore, even if they may privately pray to Creator God.

How many more must die before we reject the Evolution Religion completely?

How refreshing it is to have a new American President that **DOES** invoke God, encourages people to pray with and for him - and classy First Lady unafraid to lead a large public meeting in The Lord's Prayer while 'the Media' bitterly snipes at both of them.

Creator God has been through millennia past, is in today's totally lost world and will always be mankind's only Hope.

ALL - yes, every last bit - of the valid evidence and trustworthy explanations of that evidence are on the side of Creator God.

The Bible is God's Word and instruction Manual and is the Foundation upon which America was established.

May God forgive us for ever doubting Him and heal our land quickly!

Further, may His Only Son - yes, the Lord Jesus Christ - soon Return to be the King of kings and Lord of lords of this world forever!

May every knee bow and every tongue confess that Jesus Christ is Lord! Amen!!

List of Resources:

The Book of books:

The Holy Bible

Authored by Creator God, the Lord Jesus Christ and the
Holy Spirit - Three; yet One

Read or Listen to the Bible:

Fewer total words than most Sunday Newspapers

If listening - one can hear the entire Bible in about 40 hours

Books by men that have been critical to my YEC understanding:

The Lie - by Ken Ham - opened my eyes to young Earth Creation

Buried Alive - by Jack Cuozzo - amazing analysis of 'human' fossils

Darwin's Black Box - excellent discussion of irreducible complexity

The Genesis Flood - by Dr. Henry Morris - foundational knowledge

Tornado in a Junkyard - by James Perloff - absolute MUST read

In the Beginning - by Dr. Walt Brown - checkmates Evolutionists

Its a Young World After All - by Paul Ackerman - Natural 'Clocks'

Noah's Ark: A Feasibility Study - by John Woodmorappe - Get it!

The Revised Quote Book - Creation Science Foundation - Get it!

Who Broke the Baby? - by Dr. Jean Garton - Changed me forever!

The Case for a Creator - Lee Strobel - Straight to the point

All my books:

<u>Behold Now Behemoth: Dinosaurs All Over the Bible!</u> - analyzing the Hebrew words for animals brings MANY surprises
Cover artist: Jose Pujals

<u>Lumpy Oatmeal With Raisins and Cinnamon!</u> - help understanding miracles and supernatural events

<u>The Write In</u> by Erik Aldon (pen name) - a charismatic, born again, YEC gets elected President!

<u>The King James II Bible: ONLY the Words of God!</u> - for the first time in English; the Bible with NO added words of men and NO verse or chapter breaks and ALL the Words of any of the GodHead in Red - God does a LOT of talking in the Old Testament! (available ONLY electronically from me); request one for a donation to cover production and mailing at

<u>kingjames2bible@gmail.com</u>

Videos & CDs:

Dr. Kent Hovind - ALL OF THEM - just google him

God's Not Dead - shows real reason many Atheists reject God

What We Believe - by Frank Paretti CD from Focus on the Family

Websites:

AnswersInGenesis.org - treasure trove of Truth

2Peter3.com - Dr. Kent Hovind - aka Dr. Dino

Ministries:

Answers In Genesis

Institute for Creation Research

Creation Science Foundation

Creation Science Evangelism

Faith Builder's International - my organization - email:

kingjames2bible@gmail.com

Report Detailing Alaskan Finds and Carbon Dating of the Artifacts - search for:

QUATERNARY STRATIGRAPHIC NOMENCLATURE, CENTRAL ALASKA

TABLE 4 - Radiocarbon dates from central Alaska